T0136424

Industrial Artificial Intelligence Technologies and Applications

RIVER PUBLISHERS SERIES IN COMMUNICATIONS AND NETWORKING

Series Editors

ABBAS JAMALIPOUR
The University of Sydney
Australia

MARINA RUGGIERI
University of Rome Tor Vergata
Italy

The "River Publishers Series in Communications and Networking" is a series of comprehensive academic and professional books which focus on communication and network systems. Topics range from the theory and use of systems involving all terminals, computers, and information processors to wired and wireless networks and network layouts, protocols, architectures, and implementations. Also covered are developments stemming from new market demands in systems, products, and technologies such as personal communications services, multimedia systems, enterprise networks, and optical communications.

The series includes research monographs, edited volumes, handbooks and textbooks, providing professionals, researchers, educators, and advanced students in the field with an invaluable insight into the latest research and developments.

Topics included in this series include:

- Communication theory
- Multimedia systems
- Network architecture
- Optical communications
- Personal communication services
- Telecoms networks
- Wi-Fi network protocols

For a list of other books in this series, visit www.riverpublishers.com

Industrial Artificial Intelligence Technologies and Applications

Editors

Ovidiu Vermesan
SINTEF, Norway

Franz Wotawa
TU Graz, Austria

Mario Diaz Nava
STMicroelectronics, France

Björn Debaillie
imec, Belgium

Routledge
Taylor & Francis Group

LONDON AND NEW YORK

Published 2022 by River Publishers
River Publishers
Alsbjergvej 10, 9260 Gistrup, Denmark
www.riverpublishers.com

Distributed exclusively by Routledge
4 Park Square, Milton Park, Abingdon, Oxon OX14 4RN
605 Third Avenue, New York, NY 10017, USA

Industrial Artificial Intelligence Technologies and Applications / by Ovidiu Vermesan, Franz Wotawa, Mario Diaz Nava, Björn Debaillie.

ISBN: 978-87-7022-791-9 (hardback)
978-10-0085-203-5 (online)
978-10-0337-738-2 (master ebook)
DOI: 10.1201/9781003377382

Dedication

"Without change there is no innovation, creativity, or incentive for improvement. Those who initiate change will have a better opportunity to manage the change that is inevitable."

- William Pollard

"The brain is like a muscle. When it is in use we feel very good. Understanding is joyous."

- Carl Sagan

"By far, the greatest danger of Artificial Intelligence is that people conclude too early that they understand it."

- Eliezer Yudkowsky

Acknowledgement

The editors would like to thank all the contributors for their support in the planning and preparation of this book. The recommendations and opinions expressed in the book are those of the editors, authors, and contributors and do not necessarily represent those of any organizations, employers, or companies.

Ovidiu Vermesan
Franz Wotawa
Mario Diaz Nava
Björn Debaillie

Contents

8   Feasibility of Wafer Exchange for European Edge AI Pilot Lines   103

Annika Franziska Wandesleben, Delphine Truffier-Boutry,
Varvara Brackmann, Benjamin Lilienthal-Uhlig, Manoj Jaysnkar,
Stephan Beckx, Ivan Madarevic, Audde Demarest, Bernd Hintze,
Franck Hochschulz, Yannick Le Tiec, Alessio Spessot,
and Fabrice Nemouchi

9   A Framework for Integrating Automated Diagnosis into Simulation   113

David Kaufmann and Franz Wotawa

10   Deploying a Convolutional Neural Network on Edge MCU and Neuromorphic Hardware Platforms   129

Simon Narduzzi, Dorvan Favre, Nuria Pazos Escudero
and L. Andrea Dunbar

Preface

Industrial Artificial Intelligence Technologies and Applications

Digitalisation and Industry 5.0 are changing how manufacturing facilities operate by deploying many sensors/actuators, edge computing, and IIoT devices and forming intelligent networks of collaborative machines that are able to collect, aggregate, and intelligently process data at a network's edge.

Given the vast amount of data produced by IIoT devices, computing at the edge is required. In this context, edge computing plays an important role – the edge should provide computing resources for edge intelligence with dependability, data management, and aggregation provision in mind. Edge intelligence – for example, AI technologies with edge computing for training/learning, testing, or inference – is essential for IIoT applications to build models that can learn from a large amount of aggregated data.

Edge computing is a distributed computing paradigm that brings computation and data storage closer to a device's location. AI algorithms process the data created on a device with or without an internet connection. These new AI-based algorithms allow data to be processed within a few milliseconds, providing real-time feedback.

The AI models operate on the devices themselves without the need for a cloud connection and without the problems associated with data latency, which results in much faster data processing and support for use cases that require real-time inferencing.

Major challenges remain in achieving this potential due to the inherent complexity of designing and deploying energy-efficient edge AI algorithms and architectures, the intricacy of complex variations in neural network architectures, and the underlying limited processing capabilities of edge AI accelerators.

Industrial-edge AI can run on various hardware platforms, from ordinary microcontrollers (MCUs) to advanced neural processing devices. IIoT edge AI-connected devices use embedded algorithms to monitor device behaviour and collect and process device data. Devices make decisions, automatically correct problems, and predict future performance.

AI-based technologies are used across industries by introducing intelligent techniques, including machine and deep learning, cognitive computing, and computer vision. The application of the techniques and methods of AI in the industrial sector is a crucial reference source that provides vital research on implementing advanced technological techniques in this sector.

This book offers comprehensive coverage of the topics presented at the "International Workshop on Edge Artificial Intelligence for Industrial Applications (EAI4IA)" in Vienna, 25-26 July 2022. EAI4IA is co-located with the 31st International Joint Conference on Artificial Intelligence and the 23rd European Conference on Artificial Intelligence (IJCAI-ECAI 2022). It combines the ideas and concepts developed by researchers and practitioners working on providing edge AI methods, techniques, and tools for use in industrial applications.

By highlighting important topics, such as embedded AI for semiconductor manufacturing and trustworthy, dependable, and explainable AI for the digitising industry, verification, validation and benchmarking of AI systems and technologies, AI model development workflows and hardware target platforms deployment, the book explores the challenges faced by AI technologies deployed in various industrial application domains.

The book is ideally structured and designed for researchers, developers, managers, academics, analysts, post-graduate students, and practitioners seeking current research on the involvement of industrial-edge AI. It combines the latest methodologies, tools, and techniques related to AI and IIoT in a joint volume to build insight into their sustainable deployment in various industrial sectors.

The book is structured around four different topics:

1. **Verification, Validation and Benchmarking of AI Systems and Technologies.**
2. **Trustworthy, Dependable AI for Digitising Industry.**
3. **Embedded AI for semiconductor manufacturing.**
4. **AI model development workflow and HW target platforms deployment.**

In the following, the papers published in this book are briefly discussed.

S. Narduzzi, L. Mateu, P. Jokic, E. Azarkhish, and A. Dunbar: "Benchmarking Neuromorphic Computing for Inference" tackle the challenge of benchmarking aiming at providing a fair and user-friendly method. The authors introduce the challenge and finally come up with possible key performance indicators.

M. Molendijk, K. Vadivel, F. Corradi, G-J. van Schaik, A. Yousefzadeh, and H. Corporaal: "Benchmarking the Epiphany Processor as a Reference Neuromorphic Architecture" compare different implementations of neuromorphic processors and present suggestions for improvements.

P. Vijayan, A. Yousefzadeh, M. Sifalakis, and R. van Leuken: "Temporal Delta Layer: Exploiting Temporal Sparsity in Deep Neural Networks for Time-Series Data" deal with improving the learning of time-series data in the context of deep neural networks. In particular, the authors consider sparsity and show experimentally overall improvements.

D. Purice, M. Ludwig, and C. Lenz: "An End-to-End AI-based Automated Process for Semiconductor Device Parameter Extraction" present a validation pipeline aiming at gaining trust in semiconductor devices relying on authenticity checking. The authors further evaluate their approach by considering several artificial neural network architectures.

D. Morits, M. Rizzo Piton, and T. Laakko: "AI machine vision system for wafer defect detection" discuss the use of machine learning for fault detection based on images in the context of semiconductor manufacturing.

S. Al-Baddai and J. Papadoudis: "Failure detection in silicon package" discuss the use of machine learning techniques for wire-bonding inspection occurring during the packaging of semiconductors. The authors report on the accuracy of failure detection using machine learning in the complex industrial environment.

X. L. Liu, Eileen Salhofer, A. Safont Andreu, and R. Kern: "S2ORC-SemiCause: Annotating and analysing causality in the semiconductor domain" introduce a benchmark dataset to be used in the context of cause-effect reasoning for extracting causal relations.

A. Wandesleben, D. Truffier-Boutry, V. Brackmann, B. Lilienthal-Uhlig, M. Jaysnkar, S. Beckx, I. Madarevic, A. Demarest, B. Hintze, F. Hochschulz, Y. Le Tiec, A. Spessot, and F. Nemouchi: "Feasibility of wafer exchange for European Edge AI pilot lines" focus on contamination monitoring for allowing to exchange wafers among different facilities. In particular, the authors presented an analysis of whether such an exchange would be feasible considering three European research institutes.

D. Kaufmann and F. Wotawa: "A framework for integrating automated diagnosis into simulation" discuss a framework that allows the integration of model-based diagnosis algorithms in physical simulation. The framework can be used for verifying and validating diagnosis implementations for cyber-physical systems.

S. Narduzzi, D. Favre, N. Pazos Escudero, and A. Dunbar: "Deploying a Convolutional Neural Network on Edge MCU and Neuromorphic Hardware Platforms" discuss the deployment of neural networks for edge computing considering different platforms. The authors also report on the perceived effort of deployment for each of the platforms.

R. Prokscha, M. Schneider, and A. Höß: "Efficient Edge Deployment Demonstrated on YOLOv5 and Coral Edge TPU"consider the question of deployment of machine learning on the edge.

O. Vermesan and M. Coppola: "Embedded Edge Intelligent Processing for End-To-End Predictive Maintenance in Industrial Applications" presented the use of machine learning for edge computing supporting predictive maintenance using different technologies, workflows, and datasets.

L. A. Steffenel, A. Langlet, L. Hollard, L. Mohimont, N. Gaveau, M. Copola, C. Pierlot, and M. Rondeau: "AI-Driven Strategies to Implement a Grapevine Downy Mildew Warning System" outline the use of machine learning for identifying infections occurring in vineyards and present an experimental evaluation comparing different machine learning algorithms.

F. Wotawa and O. Tazl: "On the Verification of Diagnosis Models" focus on challenges of verification and in particular testing applied to logic-based diagnosis. The authors consider testing system models and use a running example for demonstrating how such models can be tested and come up with open research questions.

List of Figures

List of Tables

List of Contributors

Al-Baddai, Saad, *Infineon Technologies AG, Germany*

Andreu, Anna Safont, *University of Klagenfurt, Austria, Infineon Technologies Austria*

Azarkhish, Erfan, *CSEM, Switzerland*

Beckx, Stephan, *imec, Belgium*

Brackmann, Varvara, *Fraunhofer IPMS CNT, Germany*

Coppola, Marcello, *STMicroelectronics, France*

Corporaal, Henk, *Technical University of Eindhoven, Netherlands*

Corradi, Federico, *imec, Netherlands*

Demarest, Audde, *Université Grenoble Alpes, CEA-Leti, France*

Dunbar, Andrea, *CSEM, Switzerland*

Escudero, Nuria Pazos, *HE-Arc, Switzerland*

Favre, Dorvan, *CSEM, Switzerland, HE-Arc, Switzerland*

Gaveau, Nathalie, *Université de Reims Champagne Ardenne, France*

Höß, Alfred, *Ostbayerische Technische Hochschule Amberg-Weiden, Germany*

Hintze, Bernd, *FMD, Germany*

Hochschulz, Franck, *Fraunhofer IMS, Germany*

Hollard, Lilian, *Université de Reims Champagne Ardenne, France*

Jaysnkar, Manoj, *imec, Belgium*

Jokic, Petar, *CSEM, Switzerland*

Kaufmann, David, *Graz University of Technology, Austria*

Kern, Roman, *Graz University of Technology, Austria*

Laakko, Timo, *VTT Technical Research Centre of Finland Ltd, Finland*

Langlet, Axel, *Université de Reims Champagne Ardenne, France*

Le Tiec, Yannick, *Université Grenoble Alpes, CEA, LETI, France*

Lenz, Claus, *Cognition Factory GmbH, Germany*

Leuken, Rene van, *TU Delft, Netherlands*

Lilienthal-Uhlig, Benjamin, *Fraunhofer IPMS CNT, Germany*

Liu, Xing Lan, *Know-Center GmbH, Austria*

Ludwig, Matthias, *Infineon Technologies AG, Germany*

Madarevic, Ivan, *imec, Belgium*

Mateu, Loreto, *Fraunhofer IIS, Germany*

Mohimont, Lucas, *Université de Reims Champagne Ardenne, France*

Molendijk, Maarten, *imec, Netherlands, Technical University of Eindhoven, Netherlands*

Morits, Dmitry, *VTT Technical Research Centre of Finland Ltd, Finland*

Narduzzi, Simon, *CSEM, Switzerland*

Nemouchi, Fabrice, *Université Grenoble Alpes, CEA, LETI, France*

Papadoudis, Jan, *Infineon Technologies AG, Germany*

Pierlot, Clément, *Vranken-Pommery Monopole, France*

Piton, Marcelo Rizzo, *VTT Technical Research Centre of Finland Ltd, Finland*

Prokscha, Ruben, *Ostbayerische Technische Hochschule Amberg-Weiden, Germany*

Purice, Dinu, *Cognition Factory GmbH, Germany*

Rondeau, Marine, *Vranken-Pommery Monopole, Reims, France*

Salhofer, Eileen, *Know-Center GmbH, Austria, Graz University of Technology, Austria*

Schneider, Mathias, *Ostbayerische Technische Hochschule Amberg-Weiden, Germany*

Sifalakis, Manolis, *imec, Netherlands*

Spessot, Alessio, *imec, Belgium*

Steffenel, Luiz Angelo, *Université de Reims Champagne Ardenne, France*

Tazl, Oliver, *Graz University of Technology, Austria*

Truffier-Boutry, Delphine, *Université Grenoble Alpes, CEA, LETI, France*

Vadivel, Kanishkan, *Technical University of Eindhoven, Netherlands*

van Schaik, Gert-Jan, *imec, Netherlands*

Vermesan, Ovidiu, *SINTEF AS, Norway*

Vijayan, Preetha, *TU Delft, Netherlands, imec, Netherlands*

Wandesleben, Annika Franziska, *Fraunhofer IPMS CNT, Germany*

Wotawa, Franz, *Graz University of Technology, Austria*

Yousefzadeh, Amirreza, *imec, Netherlands*

1

Benchmarking Neuromorphic Computing for Inference

Simon Narduzzi[1], Loreto Mateu[2], Petar Jokic[1], Erfan Azarkhish[1], and Andrea Dunbar[1]

[1]CSEM, Switzerland
[2]Fraunhofer IIS, Germany

Abstract

In the last decade, there has been significant progress in the IoT domain due to the advances in the accuracy of neural networks and the industrialization of efficient neural network accelerator ASICs. However, intelligent devices will need to be omnipresent to create a seamless consumer experience. To make this a reality, further progress is still needed in the low-power embedded machine learning domain. Neuromorphic computing is a technology suited to such low-power intelligent sensing. However, neuromorphic computing is hampered today by the fragmentation of the hardware providers and the difficulty of embedding and comparing the algorithms' performance. The lack of standard key performance indicators spanning across the hardware-software domains makes it difficult to benchmark different solutions for a given application on a fair basis. In this paper, we summarize the current benchmarking solutions used in both hardware and software for neuromorphic systems, which are in general applicable to low-power systems. We then discuss the challenges in creating a fair and user-friendly method to benchmark such systems, before suggesting a clear methodology that includes possible key performance indicators.

Keywords: neuromorphic, inference, accelerators, benchmarking, low power, IoT, ASIC, key performance indicators.

1

DOI: 10.1201/9781003377382-1

1.1 Introduction

The performance necessary for consumer uptake of IoT devices has not been achieved yet. Intelligent always-on edge devices and sensors powered by AI and running on ultra-low power devices require outstanding energy efficiencies, low latency (real-time), high-throughput, and uncompromised accuracy. Neuromorphic computing rises to the challenge; however, the neuromorphic computing landscape is fragmented with no universal Key Performance Indicators (KPI), and comparison on a fair basis remains illusive [1]. The landscape is complex: comparisons should consider various aspects such as industrial maturity, CMOS technology implications, arithmetic precision, silicon area, power consumption, and accuracy obtained from neural networks running on the devices. Comparing target use-cases has the advantage of looking at the system-wide requirements but adds additional complexity. For example, if we take into account the inference frequency, this affects the current leakage and active power, significantly impacting the mean power consumption of the system.

The most commonly accepted quantitative metrics for benchmarking neuromorphic hardware are TOPS (Tera Operations Per Second) for throughput, TOPS/W for energy efficiency, and TOPS/mm2 for area efficiency. Hardware metrics rarely take into account the algorithmic structure. For software, the performance of Machine Learning (ML) algorithms is usually defined for a given task. Their KPIs generally target the prediction performance in terms of reached objective (often accuracy). Until recently, the KPIs rarely accounted for algorithm complexity, the computational cost, or the structure which impacts its performance on a given hardware.

Moreover, these metrics are only applicable to traditional neural networks, such as Deep Neural Network (DNNs), while for Spiking Neural Networks (SNN), other metrics such as energy per synaptic operation for energy efficiency are used. Indeed, the very nature of these DNNs and SNNs prohibits a comparison based on standard NN parameters.

The main questions asked by end-users, system integrators, and sensor manufacturers are: what is the best solution for the application, and whether a given neuromorphic processor provides some advantages over the state-of-art microcontrollers. The inability to answer these questions thwarts the industrial interest. This white paper provides a brief guide to relevant metrics for fair benchmarking of neuromorphic inference accelerator ASICs, aiming to help compare different hardware approaches for various use-cases.

The paper is organized as follows: Section 1.2 provides an overview of the state-of-the-art benchmarking of inference accelerators at algorithm and hardware levels. Then we look specifically at the KPIs which are applicable to neuromorphic or power-sensitive applications, explaining what influences the metrics. Section 1.3 explains why combining KPIs for both hardware and algorithms is essential for fair benchmarking of neuromorphic computing. Finally, Section 1.4 summarizes and concludes the paper.

1.2 State-of-the-art in Benchmarking

Benchmarking of NNs inference performance for a task occurs at both the algorithm and hardware levels. The use-case provides the constraints and optimizations to be achieved through the combination of the ML model and the hardware. Currently, ML algorithms and hardware are usually benchmarked independently with their own metrics.

For ML algorithms, task-related metrics are the standard. Usually, the task-related metrics are independent of the nature of the ML model used, allowing the comparison between the algorithmic techniques used to perform the task: while the algorithm may change, the way to assess the performance of the algorithm on a certain task (e.g., image classification) remains the same. This methodology allows rapid development of deep learning techniques by comparing the performance of the algorithms on a given task. In order to target resource-limited IoT applications, metrics measuring the complexity of the model exist, such as the number of parameters, sparsity, depth, and (floating-point) operation counts, are taken into account. These KPIs are measurable via simulation of the model, and most of the current deep learning libraries now provide functions that report these KPIs.

On the other hand, hardware KPIs are extracted from the deployment platform while running a certain algorithmic model. They can be either simulated or computed by running the target application on the device. These KPIs usually include power consumption (estimation), latency, and memory metrics. In other words, they provide performance results of an ML algorithm for a certain use case on a specific hardware platform. This gives a good representation of how a single device works for a given use-case but makes benchmarking difficult. In the following sections, we present the current state-of-the-art solutions to benchmarking software and hardware with a focus on low-power devices. A summary of the standard KPIs is given in Table 1.1.

Table 1.1   Relevant KPIs for tasks, models and hardware domains. We also mention some combined KPIs to illustrate the inter-dependency of the domains.

	Metric	Definition	Unit
Task KPIs	Objective function	Determines and measures the goal of a given task	-
	(Balanced) Accuracy	Computes the ratio of correctly classified examples over the dataset (weighted by class occurrences)	%
	Precision	Computes the ratio of correctly classified samples per class	%
	True Positive Rate (TPR) / Sensitivity / Recall	Ratio of true positives over the total number of samples	%
	True Negative Rate (TNR) / Specificity	Ratio of true negatives over the total number of samples	%
	False Positive Rate (FPR)	Ratio of false positives over the total number of samples	%
	Mean average precision (mAP)	The mean of the average precision per sample	%
	F1-Score	Harmonic mean of the precision and recall	-
	Receiver operating characteristic (ROC)	Plot the true positive rate against the false positive rate of a class	-
	Area under the curve (AOC)	The area under the ROC curve	-
	Metric	**Definition**	**Unit**
Model KPIs	Complexity	Number of multiply-accumulate (multiply-additions) operations	MACs (MAdds)
		Number of (floating-point) operations, often assumed equivalent to 2x MACs	(FL)OPs
	Parameters	Total number of parameters in the model (weights, biases, etc.)	-
	Precision	Precision of the parameters and activations (floating point, integer, etc.)	bits
	Structure	Number and type of layers (and neurons)	-
	Sparsity	Ratio of sparse values	%
	Maximum Activation	Maximum activation buffer of any layer in the network	-
	Spike Count	Number of spikes emitted	-
	Spike rate / Spike Frequency	Rate at which the spikes are emitted (model level or neuron level)	Hz
	SynOps	Number of synaptic operations produced by the model at inference	-
	Metric	**Definition**	**Unit**
Hardware KPIs	(Idle/Peak) Power consumption	Power consumption of the system in idle/peak mode	Watts (W)
	(Peak) Number of operations	Number of operations per second (peak over short period of time)	TOPS (TOPS/s)
	Die size	Silicon area of the system	mm^2
	Memory size	On chip memory size	MB
	Memory bandwidth	Maximum data rate from memory (internal/external)	bit/s
	Energy per operation	Energy to perform one operation including necessary memory transfers	J/op
	Energy per synaptic operations	Energy to perform one synaptic operation including necessary memory transfers	J/SynOps
	Mean energy per spike	Energy required to execute one spike	J/Spike
	Energy efficiency	Tera operations per second per watt	TOPS/W
	Area efficiency	Tera operations per mm^2	$TOPS/mm^2$
	Precision	Precision of the parameters and activations (floating point, integer, etc.)	bits
	Maximum network size	Size constraints of network architecture (number of neurons/synapses, etc.)	
	Max. Core Frequency	Maximum frequency of the hardware	MHz
	Metric	**Definition**	**Unit**
Combined	Wake-up time	Time to load the model in memory and make it ready for inference	ms
	Latency	Time needed to perform inference	ms
	Inference rate / Throughput	Number of inferences per second	Inference/s
	Energy per inference	Energy consumed by the system when running an inference for a certain model	Joules (J)

1.2.1 Machine Learning

Machine learning techniques, and especially deep learning algorithms, are engineered iteratively for a given task's performance. ML algorithms are typically compared in terms of accuracy for a given task, such as segmentation or classification on a specified dataset. The task performance comparison is nowadays well established in the ML community. For classification tasks, accuracy, precision, recall, receiver operating characteristics (ROC), and area under the curve (AUC) are some of the most frequently used metrics. A typical example of a table is shown in Table 1.2. We refer the reader to [2, 3, 4] for a more detailed overview of relevant metrics in ML tasks.

In order to give fair comparison for different domains of deep learning, training and test datasets have been established. According to PapersWith-Code [6], computer vision-related tasks have the largest number of datasets, with long-established quasi-standards such as CIFAR [7], ImageNet [8], and COCO [5]. Specific computer vision tasks have their own standard datasets, such as KITTI [10] for autonomous driving and FDDB [11] and WIDER Face [12] for face detection applications. Natural Language Processing (NLP) tasks are the second most popular tasks for machine learning, with near 2000 datasets comprising GLUE [13] and SQuAD [14] benchmarks. Audio, biomedical and physics-related tasks equally have their own datasets. It should be mentioned that other ML techniques also have their own equivalent dataset for example reinforcement learning (RL) tasks also have their own standard benchmarks e.g. OpenAI Gym [15] which contains a set of tasks to test reinforcement learning algorithms. Here the tasks take place in a virtual environment, and all the physics and interactions are handled by the environment.

Table 1.2 Accuracy (*Acc*) for different object detection settings on COCO test-dev. Adapted from [9].

Model	Acc	Acc_{50}	Acc_{75}	Acc_S	Acc_M	Acc_L
YOLOv2	21.6	44.0	19.2	5.0	22.4	35.5
SSD513	31.2	50.4	33.3	10.2	34.5	49.8
DSSD513	33.2	53.3	35.2	13.0	35.4	51.1
RetinaNet (ours)	39.1	59.1	42.3	21.8	42.7	50.2

The importance of the data set

The importance of the datasets can clearly be seen when looking at SNNs. Currently, the performance of SNNs does not reach DNN performance. Research in SNNs has focused on the structure of the network and learning algorithms rather than on task performance. Thus, the work used well-known datasets for DNNs and transformed them into event-based versions, such as MNIST-DVS, N-MNIST, and N-Caltech101[16]. Only recently, with the technology of event-based cameras, have SNN been applied to adapted datasets for various use-cases (e.g., DVS128[17] and TIDIGITS[18]). These new datasets will now allow us to see if SNNs can truly rival their DNN counterparts.

The standard ML benchmarking, as discussed above, usually focuses on accuracy. This means that the resources needed due to the underlying algorithm complexity, and thus power consumption, are ignored. In resource-constrained use cases such as those in edge ML, the models are designed to provide a computational advantage. For resource-constrained systems assessing the algorithmic performance on a target task, algorithms can be compared in terms of complexity, which determines the runtime constraints. In classical machine learning, there are well-established metrics for comparing the complexity of algorithms. For example, decision trees are defined by the number of nodes and depth of the tree [19]. NNs, on the other hand, are usually compared in terms of number of parameters or number of MAC operations [20, 21, 22]. We refer the reader to the survey by Hu et al. [23] for further discussion about model complexity. Table 1.3 shows a classic representation of results for an edge ML algorithm, taking into account the resources used:

In low power systems, the number of operations, multiply-accumulate (MAC), or multiply-add (MAD) are also used as an NN optimization parameter. The computation latency of an arithmetic block is also highly dependent

Table 1.3   Representation of resource-constrained KPIs, adapted from [20].

Network	mAP	Params	MAdds	CPU inference time
SSD300	23.2	36.1M	35.2B	-
SSD512	26.8	36.1M	99.5B	-
YOLOv2	21.6	50.7M	17.5B	-
MNetV1+SSDLite	22.2	5.1M	1.3B	270ms
MNetV2+SSDLite	22.1	4.3M	0.8B	200ms

on the precision used to represent the weights and activation of the NN (i.e., 8bit computations usually run at higher frequencies than for 32bits). For tiny devices, the type and number of layers of neural networks may be a metric of interest, as some hardware may be optimized for certain architectures: some platforms support separable convolutions, while others do not. The maximum supported activation size for a network layer can also be a limiting factor since some models might exceed this constraint for some embedded platforms.

Standard SNN topologies have also been compared using frameworks [24]. Among the metrics that can be used to compare SNN models, the type of neurons and synapses, the number of emitted spikes and synaptic operations, and the rate of the SNNs are the most often used.

It remains difficult, however, to compare cross-paradigm algorithms, especially when comparing deep learning with emerging paradigms like SNNs. While some efforts have been made to compare ANN and SNNs [25], a standard set of metrics has still to be defined.

1.2.2 Hardware

An increasing number of hardware evaluation tools aim at benchmarking ML applications directly on the hardware. For example, QuTiBench [37] presents a benchmarking tool that takes algorithmic optimization and co-design into account. The MLMark[27] benchmark targets ML applications running on MCUs at the edge. However, both QuTiBench and MLMark models are too large for tiny applications and require large memories, which are not available on tiny edge devices. TinyMLPerf [28] provides benchmarks for tiny systems based on imposed models and tasks, yielding the latency and speed-related KPIs. Submission of results using other network architectures is allowed in its open division. Further tools, like SMAUG [29], MAESTRO[30] and Aladdin[31], provide software solutions to emulate workloads on deep-learning accelerators using varying topologies.

The power consumption of edge ML processing hardware is of utmost interest as it directly impacts the battery lifetime of a system. Dynamic power dominates in most high-throughput applications, while leakage power is only significant in low duty cycle modes[32], where power gating, body biasing, and voltage scaling techniques are employed to reduce leakage. Peak power consumption corresponds to the maximum power consumption measured, which becomes relevant for battery- or energy harvesting-supplied applications.

The throughput metric indicates the number of operations that the hardware can perform per second, while latency is the time needed to perform an entire inference. Note that the peak throughput can usually not be reached for all network topologies, and latency does not directly scale with parallelization, as the peak throughput does[33]. Thus, latency is a combined HW/SW metric. It can be measured by running multiple inferences and afterward averaging the execution time. All parameters to run the inference should be loaded before measuring the inference time.

The CMOS technology employed for the hardware design impacts the die size and the area efficiency, and thus also directly determines its cost. Area efficiency provides a figure of merit between the throughput, limited by hardware resources and frequency, that can be achieved per area. On-chip memory size provides a raw estimation of the number of parameters of the NN that can be stored on the chip. In a multi-core architecture, usually, both the number of neurons and number of synapses per core are given.

Energy efficiency refers to the throughput that can be achieved per watt, which is equivalent to the number of operations per Joule. For obtaining this KPI, a NN is deployed to an inference accelerator, while execution time and power consumption are measured for performing inference. In the case of NNs, the multiply and accumulate (MAC) operation corresponds to two operations. Note that the bit precision of each operation directly impacts both the accuracy and the energy efficiency (e.g., 32bits float versus 8bits integer) and must therefore be carefully traded off. Energy per operation and energy per neuron are fair metrics if the bit resolution is provided since they are independent of the NN algorithm employed and therefore only hardware-related.

Some hardware only supports a limited number of layers and layer types with restricted dimensions. Others provide optimizations and specialized units. These optimizations, while not being directly comparable, have a strong impact on the hardware KPIs. Furthermore, power consumption is influenced by the core voltage supply, which depends on the CMOS technology used for the hardware design. Thus, the energy efficiency metric (TOPS/W) can be misleading unless all hardware restrictions are known. The same applies to other representations like GOPS/W. Typical display of performance in terms of OPS and associated power are presented in Table 1.4. and from these

Table 1.4 Typical display of performance comparison of neuromorphic hardware platforms, adapted from [34].

Accelerator	Type	Target application	Performance
NVIDIA Jetson Nano	GPU	Embedded	472 GOPS @ 5 – 10 W
Nvidia Jetson TX2	GPU	Edge	1,3 TOPS @ 7,5 W
NVIDIA Jetson AGX Xavier	GPU	Edge	30 TOPS @ 30 W
NVIDIA Drive AGX Pegasus	GPU	Automotive	320 TOPS
Intel Movidius Myriad 2 bzw. Myriad X	Chip	Embedded/Edge DL/Vision	4 TOPS @ 1 W (Myriad X)
MobilEye EyeQ4	Chip	Automotive	2.5 TOPS @ 3 W
GreenWaves GAP8	Chip	Battery powered AI	200 MOPS bis 8 GOPS @ <100mW
Canaan Kendryte K210	Chip	Embedded Vision & Audio	250 GOPS @ 300mW
Google Coral Edge TPU	Chip	Edge	4 TOPS @ <2,5W
Lattice sensAI Stack	Soft IP-Core	Embedded	<1 mW – 1 W
Videantis v-MP6000UDXM	Soft IP-Core	Embedded DL/Vision	<6,6 TOPS @ 400 MHz

Table 1.5 Recent display of performance comparison of neuromorphic hardware platforms, adapted from [35].

		Eyeriss	ENVISION	Thinker	UNPU	This work	
Technology		65nm	28nm	65nm	65nm	65nm	
Area		1176k gates (NAND-2)	1950k gates (NAND-2)	2950k gates (NAND-2)	4.0mm×4.0mm (Die Area)	2695k gates (NAND-2)	
On-chip SRAM (kB)		181.5	144	348	256	246	
Max Core Frequency (MHz)		200	200	200	200	200	
Bit Precision		16b	4b/8b/16b	8b/16b	1b-16b	8b	
Num. of MACs		168 (16b)	512 (8b)	1024 (8b)	13824 (bit-serial)	384 (8b)	
DNN Model		AlexNet	AlexNet	AlexNet	AlexNet	sparse AlexNet	sparse MobileNet
Batch Size		4	N/A	15	N/A	1	1
Core Frequency (MHz)		200	200	200	200	200	200
Bit Precision		16b	N/A	adaptive	8b	8b	8b
Inference/sec	(CONV only)	34.7	47	-	346	342.4	-
	(Overall)	-	-	254.3	-	278.7	1470.6
Inference/J	(CONV only)	124.8	1068.2	-	1097.5	743.4	-
	(Overall)	-	-	876.6	-	664.6	2560.3

terms the TOPS/W metric can be extrapolated. However, recent publications provide combined metrics as it is shown in Table 1.5.

Processing hardware is limited by the supported arithmetic precisions for parameters and activations, with the previously mentioned effects on accuracy. Some hardware implementations allow for several bit resolutions, allowing to dynamically trade-off throughput, memory needs, and accuracy. Generally, lower precisions lead to lower algorithmic accuracy.

1.3 Guidelines

Benchmarking of ML applications cannot be tackled as a standalone problem at the level of either only hardware or algorithms. A holistic view requires a wide range of expertise and domains. It requires a multidisciplinary and multidimensional approach considering, among other things, the hardware platform, the NN (model), and the use-case under evaluation. In order to make the right choices for building blocks, the system integrator needs to know the KPIs for a given use-case that different NNs will be able to deliver on different hardware platforms.

This section explains why a multidisciplinary approach combining both algorithms and hardware is needed to avoid drawing unfair and misleading conclusions and comparisons. In the following, we first describe what is unfair and fair benchmarking in Section 1.1, and then present a combined KPI approach and guidelines for benchmarking in sections 1.2 and 1.3.

1.3.1 Fair and Unfair Benchmarking

With the new generations of hardware accelerators, many optimizations in hardware try to co-optimize energy and performance, such as zero-skipping components, in-memory computing, and multi-core convolution units. However, it is sometimes unclear if these optimization features are correctly exploited when embedding complex deep learning models. This lack of transparency in the optimizations and embedding processes of the models results in sub-optimal deployments in the hardware. Furthermore, SDK documentation for a large number of accelerators is unclear or lacks critical content for high-level developers and data scientists to perform inference-time optimizations. This makes the embedding process and the subsequent measurements of the KPIs difficult.

Today, most models deployed on hardware are trained on GPU machines and deployed on target hardware platforms using their respective optimizations. The wide variety of optimizations employed in different hardware implementations [36,37] target specific use-cases, which might favor one or the other (benchmarking) algorithms (and the underlying layer types), further complicating fair benchmarking. Thus, there are hardware solutions that outperform others by orders of magnitude for specific tasks while providing poor performance in others. This type of benchmarking is unfair, as the models are not optimized and thus do not take advantage fully of each platform. Their KPIs are comparable, but the benchmarking is unfair with respect to the hardware, as a specially designed model for a particular platform could be more performant than another model deployed on another platform, see Figure 1.1a. This shows that use-case-agnostic benchmarking can be misleading. A platform might receive a low score with general benchmarks, while performing excellently for a hardware-tailored task.

In contrast, a fair benchmarking based on a defined use-case (independently of the model used) would exploit all the tools and optimizations provided by the constructor to exploit the hardware to its full potential. However, the results of the benchmarks can be challenging to compare, as the base model and optimizations are different between the compared hardware, see Figure 1.1b. If we compare with conventional benchmarking of processors, the benchmarks do not account for the underlying optimizations; a superscalar processor will be benchmarked against a non-superscalar processor using the same tests.

One particular aspect to take into account in the design of an inference accelerator is the selection of the CMOS technology and embedded non-volatile memory (eNVM). If eNVM is used for leveraging from the lack of power consumption for retaining the stored values after writing, the qualification of the memory by the foundry in the selected CMOS process is necessary for its industrialization and therefore a crucial criterion. The selection of the CMOS process has an impact on the cost and size of the inference accelerator IP that needs to be considered. Moreover, the CMOS process has also an impact on the active power and leakage power of the inference ASIC and needs to be part of the information provided for a fair comparison between inference accelerators fabricated in different CMOS processes.

There still remain challenges in the method of comparison. Benchmarking approaches for Von-Neumann architectures are relatively widespread and standardized [38, 39]. By contrast, clear benchmarking methodologies for non-Von-Neumann architectures do not exist yet, making them difficult to compare. In particular, neuromorphic circuit design is an emerging multidisciplinary challenge that is still in an exploratory phase making the comparison of the underlying hardware difficult due to its variety. Although many existing techniques report significantly reduced energy consumption figures, they still compare themselves to standard low-power microcontrollers.

Benchmarking should be done at different stages and abstraction levels, considering various aspects such as the algorithm performance, the technical characteristics, the architectural parameters, and the flexibility and amenities hardware provides for a specific use-case. As of today, different KPI values can be obtained with the same algorithm and same hardware just by changing the use-case from always-on to event-based.

1.3.2 Combined KPIs and Approaches for Benchmarking

The application deployment KPIs are at the intersection of the performance indicators required by a given use-case, the model solving the task, and

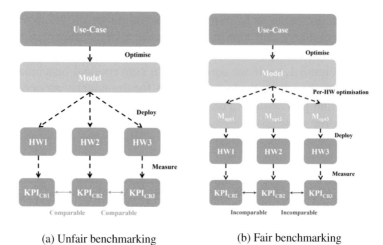

(a) Unfair benchmarking   (b) Fair benchmarking

Figure 1.1   Benchmarking fairness. (a) Unfair benchmarking: the KPIs are comparable, but the benchmarked hardware platforms are not exploited to their full potential. (b) Fair benchmarking: the hardware platforms are exploited to their full potential, but the resulting combined KPIs (KPI$_{CB}$) are not comparable.

the hardware system on which the application is deployed, see Figure 1.2. Because of the large number of KPIs that can be reported, it is difficult to have an objective comparison between different platforms, as a platform can perform well on certain KPIs and poorly on others (e.g., simulating an SNN on a CNN accelerator). Furthermore, not all platforms report the same set of metrics and the metrics are not usually convertible to each other (e.g., energy consumption is not always relying only on MAC operations).

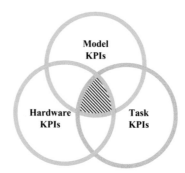

Figure 1.2   Combined KPIs for fair benchmarking

Some task-related metrics heavily depend on the use-case and application scenarios, and should be used only in these specific cases. For example, the performance of a keyword spotting algorithm should not be compared with the one of an object classification algorithm, even though both aim at high accuracy. For these reasons, a (small) set of KPIs are desirable which have the following properties:

- Orthogonality
- Reproducibility
- Objectiveness
- Use-case independence

To assess the performance of NN models running on hardware for a certain use-case, the KPIs should be combined, as shown in Table 1, to express the performance of the application on the hardware platform. In this regard, Fra et al. [40] have proposed a multi-metric approach taking into account: 1) accuracy, 2) number of parameters of the NN, 3) memory footprint in MB. These three metrics provide an overview of the NNs: which one provides better results in the classification task and which one has a smaller memory footprint. Further metrics which should now be taken into consideration are: 4) Energy consumption per inference, 5) the number of operations per second.

The resulting KPIs of the deployment could also contain an indicator about the flexibility of the hardware accelerator. For comparison in terms of flexibility, it is necessary to indicate the supported layer types, the supported bit resolution for inputs, parameters and activation functions, and the sizes of the kernel filters. By combining metrics that depend on the NN algorithm and the hardware, a fair comparison for a use-case can be achieved if the number of parameters of the NN is optimized and the dataset employed is the same.

1.3.3 Outlook : Use-case Based Benchmarking

A solution to the afore-mentioned challenge would be to propose a use-case-dependent benchmarking that does not rely at all on the model architectures of the given model. For an industrial setting, it is interesting to obtain high performance independently of the techniques used. What matters is that the application performs within the given constraints of the use-case.

A solution is illustrated in Figure 1.3. In this paradigm, a use-case would be defined by some target KPIs to reach, such as minimum accuracy

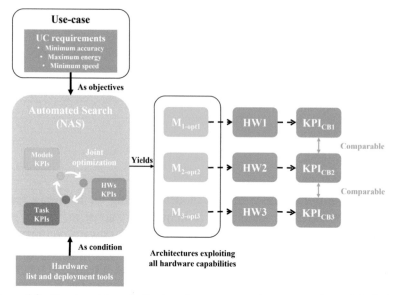

Figure 1.3   Benchmarking pipeline based on use-cases. An automated search finds the best possible model exploiting the performance offered by each target hardware platforms. The resulting combined KPIs are comparable.

and maximum energy. To benchmark the hardware, an automated search technique, such as Network Architecture Search (NAS), would try to find the model that fits the target hardware and then optimize the model further to improve the latency or memory use. This type of benchmarking would be use-case dependent and model agnostic, beside the meta-model composing the automated search. Such benchmarking method would output comparable (combined) KPIs, making the comparison of hardware and the selection of the best one possible. Of course, an extensive benchmarking suite covering several use-cases (audio-based, image-based, classification, regression, etc.) is necessary to ensure fairness across domains.

Following the methodology presented, there are some guidelines to follow in order to ensure that the extracted KPIs respect the properties presented in the section 1.2. In addition to measuring the combined KPIs, it is necessary to provide information on the entire deployment pipeline. Indeed, the KPIs related to the solved task, the (final) model deployed on the hardware, the characteristics of the hardware, and finally, the combined KPIs based on the previous information can be calculated.

The use-cases should be clearly defined and cover several machine learning tasks. Although the methodology can be applied to a single use-case to compare a few hardware platforms, the industrial application cases are generally broad. It is, therefore, preferable to select a neuromorphic platform that offers the best performance for a wide range of tasks. This can only be achieved with a benchmarking tool that is diversified in terms of the tasks to be solved.

The methodology also requires a complete software tool chain to have rapid and reproducible deployments of the NNs on the hardware. Quantization-aware training tools or even better hardware-aware training tools compatible with the target hardware platforms are beneficial. The efficient execution of algorithms does not only depend on the hardware architecture, like the processing resources, but equally on an efficient mapping strategy that schedules the hardware resources for high throughput and low power consumption. Depending on the architecture, algorithm-to-hardware compilers or on-board schedulers ensure this optimization.

Finally, adequate documentation about the hardware technology, the search algorithm used for benchmarking, the use-case realized by the benchmark, and the interpretation of the results provided by the benchmark is necessary to empower the user in its selection of the most suitable hardware platform.

1.4 Conclusion

In this paper, we have summarized the standard techniques for benchmarking NN accelerator hardware and ML software, in addition, we have specified the KPIs that are most relevant for resource aware inference. We have through example shown that, in ultra-low-power or neuromorphic systems, separating hardware and ML algorithms and use-case parameters leads to an ineffective means of comparison. Only when considering these three in a holistic manner, can system be benchmarked. Integrating KPIs that allow benchmarking at the system level in this way is complex. It is important to do this as the inability to benchmark the IoT systems today is reducing the uptake by industry. In this paper, we have proposed a benchmarking methodology based on use-cases where the ML algorithm is adapted to the hardware to allow fair comparison. Finally, we provide a guideline on what aspects are important to take into account while developing such benchmarking tool to ensure that the resulting KPIs are comparable.

Acknowledgements

This work is supported through the project ANDANTE. ANDANTE has received funding from the ECSEL Joint Undertaking (JU) under grant agreement No 876925. The JU receives support from the European Union's Horizon 2020 research and innovation programme and France, Belgium, Germany, Netherlands, Portugal, Spain, Switzerland. ANDANTE has also received funding from the German Federal Ministry of Education and Research (BMBF) under Grant No. 16MEE0116. The authors are responsible for the content of this publication.

References

[1] M. Davies. Benchmarks for progress in neuromorphic computing. *Nature Machine Intelligence*, 1(9):386–388, 2019.

[2] B. J. Erickson and F. Kitamura. Magician's corner: 9. performance metrics for machine learning models. *Radiology: Artificial Intelligence*, 3(3), 2021.

[3] A. Rácz, D. Bajusz, and K. Héberger. Multi-level comparison of machine learning classifiers and their performance metrics. *Molecules*, 24(15), 2019.

[4] F. Pedregosa, G. Varoquaux, A. Gramfort, V. Michel, B. Thirion, O. Grisel, M. Blondel, P. Prettenhofer, R. Weiss, V. Dubourg, et al. Scikit-learn: Machine learning in python. *the Journal of machine Learning research*, 12:2825–2830, 2011.

[5] T.-Y. Lin, M. Maire, S. Belongie, J. Hays, P. Perona, D. Ramanan, P. Dollár, and C. L. Zitnick. Microsoft coco: Common objects in context. In *European conference on computer vision*, pages 740–755. Springer, 2014.

[6] https://paperswithcode.com. Website, 2021.

[7] A. Krizhevsky, G. Hinton, et al. Learning multiple layers of features from tiny images. 2009.

[8] J. Deng, W. Dong, R. Socher, L.-J. Li, K. Li, and L. Fei-Fei. Imagenet: A large-scale hierarchical image database. In *2009 IEEE conference on computer vision and pattern recognition*, pages 248–255. Ieee, 2009.

[9] T.-Y. Lin, P. Goyal, R. Girshick, K. He, and P. Dollár. Focal loss for dense object detection. In *Proceedings of the IEEE international conference on computer vision*, pages 2980–2988, 2017.

[10] A. Geiger, P. Lenz, and R. Urtasun. Are we ready for autonomous driving? the kitti vision benchmark suite. In *2012 IEEE conference on computer vision and pattern recognition*, pages 3354–3361. IEEE, 2012.

[11] V. Jain and E. Learned-Miller. Fddb: A benchmark for face detection in unconstrained settings. Technical report, UMass Amherst technical report, 2010.

[12] S. Yang, P. Luo, C.-C. Loy, and X. Tang. Wider face: A face detection benchmark. In *Proceedings of the IEEE conference on computer vision and pattern recognition*, pages 5525–5533, 2016.

[13] A. Wang, A. Singh, J. Michael, F. Hill, O. Levy, and S. R. Bowman. Glue: A multi-task benchmark and analysis platform for natural language understanding. *arXiv preprint arXiv:1804.07461*, 2018.

[14] P. Rajpurkar, J. Zhang, K. Lopyrev, and P. Liang. Squad: 100,000+ questions for machine comprehension of text. *arXiv preprint arXiv:1606.05250*, 2016.

[15] G. Brockman, V. Cheung, L. Pettersson, J. Schneider, J. Schulman, J. Tang, and W. Zaremba. Openai gym. *arXiv preprint arXiv:1606.01540*, 2016.

[16] G. Orchard, A. Jayawant, G. K. Cohen, and N. Thakor. Converting static image datasets to spiking neuromorphic datasets using saccades. *Frontiers in neuroscience*, 9:437, 2015.

[17] A. Amir, B. Taba, D. Berg, T. Melano, J. McKinstry, C. Di Nolfo, T. Nayak, A. Andreopoulos, G. Garreau, M. Mendoza, et al. A low power, fully event-based gesture recognition system. In *Proceedings of the IEEE conference on computer vision and pattern recognition*, pages 7243–7252, 2017.

[18] J. Anumula, D. Neil, T. Delbruck, and S.-C. Liu. Feature representations for neuromorphic audio spike streams. *Frontiers in neuroscience*, 12:23, 2018.

[19] H. Buhrman and R. De Wolf. Complexity measures and decision tree complexity: a survey. *Theoretical Computer Science*, 288(1):21–43, 2002.

[20] M. Sandler, A. Howard, M. Zhu, A. Zhmoginov, and L.-C. Chen. Mobilenetv2: Inverted residuals and linear bottlenecks. In *Proceedings of the IEEE conference on computer vision and pattern recognition*, pages 4510–4520, 2018.

[21] N. Ma, X. Zhang, H.-T. Zheng, and J. Sun. Shufflenet v2: Practical guidelines for efficient cnn architecture design. In *Proceedings of*

the European conference on computer vision (ECCV), pages 116–131, 2018.

[22] M. Tan and Q. Le. Efficientnet: Rethinking model scaling for convolutional neural networks. In *International conference on machine learning*, pages 6105–6114. PMLR, 2019.

[23] X. Hu, L. Chu, J. Pei, W. Liu, and J. Bian. Model complexity of deep learning: A survey. *Knowledge and Information Systems*, 63(10):2585–2619, 2021.

[24] S. R. Kulkarni, M. Parsa, J. P. Mitchell, and C. D. Schuman. Benchmarking the performance of neuromorphic and spiking neural network simulators. *Neurocomputing*, 447:145–160, 2021.

[25] S. Narduzzi, S. A. Bigdeli, S.-C. Liu, and L. A. Dunbar. Optimizing the consumption of spiking neural networks with activity regularization. In *ICASSP 2022-2022 IEEE International Conference on Acoustics, Speech and Signal Processing (ICASSP)*, pages 61–65. IEEE, 2022.

[26] M. Blott. *Benchmarking Neural Networks on Heterogeneous Hardware*. PhD thesis, Trinity College, 2021.

[27] P. Torelli and M. Bangale. Measuring inference performance of machine-learning frameworks on edge-class devices with the mlmark benchmark. *Techincal Report. Available online: https://www. eembc. org/techlit/articles/MLMARK-WHITEPAPERFINAL-1. pdf (accessed on 5 April 2021)*, 2021.

[28] C. R. Banbury, V. J. Reddi, M. Lam, W. Fu, A. Fazel, J. Holleman, X. Huang, R. Hurtado, D. Kanter, A. Lokhmotov, et al. Benchmarking tinyml systems: Challenges and direction. *arXiv preprint arXiv:2003.04821*, 2020.

[29] S. Xi, Y. Yao, K. Bhardwaj, P. Whatmough, G.-Y. Wei, and D. Brooks. Smaug: End-to-end full-stack simulation infrastructure for deep learning workloads. *ACM Transactions on Architecture and Code Optimization (TACO)*, 17(4):1–26, 2020.

[30] H. Kwon, P. Chatarasi, M. Pellauer, A. Parashar, V. Sarkar, and T. Krishna. Understanding reuse, performance, and hardware cost of dnn dataflows: A data-centric approach using maestro. 2020.

[31] Y. S. Shao, B. Reagen, G.-Y. Wei, and D. Brooks. Aladdin: A pre-rtl, power-performance accelerator simulator enabling large design space exploration of customized architectures. In *2014 ACM/IEEE 41st International Symposium on Computer Architecture (ISCA)*, pages 97–108. IEEE, 2014.

[32] F. Fallah and M. Pedram. Standby and active leakage current control and minimization in cmos vlsi circuits. *IEICE transactions on electronics*, 88(4):509–519, 2005.

[33] J. Hanhirova, T. Kämäräinen, S. Seppälä, M. Siekkinen, V. Hirvisalo, and A. Ylä-Jääski. Latency and throughput characterization of convolutional neural networks for mobile computer vision. In *Proceedings of the 9th ACM Multimedia Systems Conference*, pages 204–215, 2018.

[34] M. Breiling, R. Struharik, and L. Mateu. Machine learning: Elektronenhirn 4.0. 2019.

[35] Y.-H. Chen, T.-J. Yang, J. Emer, and V. Sze. Eyeriss v2: A flexible accelerator for emerging deep neural networks on mobile devices. *IEEE Journal on Emerging and Selected Topics in Circuits and Systems*, 9(2):292–308, 2019.

[36] P. Jokic, E. Azarkhish, A. Bonetti, M. Pons, S. Emery, and L. Benini. A construction kit for efficient low power neural network accelerator designs. *arXiv preprint arXiv:2106.12810*, 2021.

[37] M. Blott, L. Halder, M. Leeser, and L. Doyle. Qutibench: Benchmarking neural networks on heterogeneous hardware. *ACM Journal on Emerging Technologies in Computing Systems (JETC)*, 15(4):1–38, 2019.

[38] EMBCC ULPMark: https://www.eembc.org/ulpmark/. Website, 2021.

[39] EMBCC CoreMark: https://www.eembc.org/coremark/. Website, 2021.

[40] V. Fra, E. Forno, R. Pignari, T. Stewart, E. Macii, and G. Urgese. Human activity recognition: suitability of a neuromorphic approach for on-edge aiot applications. *Neuromorphic Computing and Engineering*, 2022.

2

Benchmarking the Epiphany Processor as a Reference Neuromorphic Architecture

Maarten Molendijk[1,2], Kanishkan Vadivel[2], Federico Corradi[2,1], Gert-Jan van Schaik[1], Amirreza Yousefzadeh[1], and Henk Corporaal[2]

[1]imec, Netherlands
[2]Technical University of Eindhoven, Netherlands

Abstract

This short article explains why the Epiphany architecture is a proper reference for digital large-scale neuromorphic design. We compare the Epiphany architecture with several modern digital neuromorphic processors. We show the result of mapping the binary LeNet-5 neural network into few modern neuromorphic architectures and demonstrate the efficient use of memory in Epiphany. Finally, we show the results of our benchmarking experiments with Epiphany and propose a few suggestions to improve the architecture for neuromorphic applications. Epiphany can update a neuron on average in 120ns which is enough for many real-time neuromorphic applications.

Keywords: neuromorphic processor, spiking neural network, bio-inspired processing, artificial intelligence, edge AI.

2.1 Introduction and Background

Neuromorphic sensing and computing systems mimic the functions and the computational primitives of the nervous systems. Nevertheless, state-of-the-art Deep Neural Networks (DNNs) have exceeded the accuracy of biological brains (including the human brain) in specific tasks like video/audio processing, decision-making, planning and playing games. However, all of these

DOI: 10.1201/9781003377382-2

tasks are done without considering one of the main restrictions in bio-evolution, the "energy consumption". The biological restrictions pushed the evolution toward power-efficient algorithms and architectures. The human brain is an extreme example that consumes a considerable portion (around 20%) of the human body's energy while it has less than 3% of the total weight.

Even though the elements of the biological fabric in the brain are not as fast and arguably as power efficient as our modern silicon technologies, no computing platform can get close to the compute efficiency of the biological brain for processing natural signals. The brain is a perfect example of algorithm-hardware co-optimization. As mentioned, the ultimate goal of bio-inspired processing is to process the raw sensory data with the minimum amount of power consumption.

The Epiphany architecture was first introduced back in 2009 [1] as a high-performance energy-efficient many-core architecture suitable for real-time embedded systems. Epiphany's architecture contains many RISC processor cores connected with a packet-based mesh Network-On-Chip (NoC). Figure 5.1 shows the big picture of the Epiphany's architecture. This architecture is different from the mainstream von-Neumann type multi-core processors since in Epiphany, the cores are connected directly via a NoC

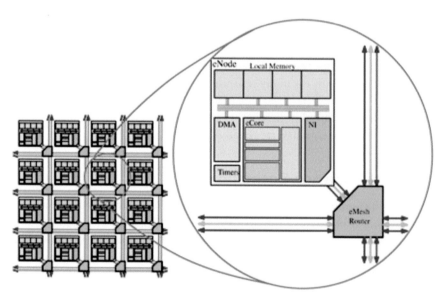

Figure 2.1   Overall scalable architecture of Epiphany-III [1].

without using a single shared memory to communicate. The mesh packet switch network in Epiphany results in highly efficient local data movement between neighbouring processors. However, it introduces a possible non-deterministic behaviour as the order of the packets in the mesh network is not guaranteed. Despite implementing a synchronization mechanism, the RISC processors work individually, and the architecture is not designed for strict synchronous execution (since it harms the scalability feature). Hence, programming epiphany with a conventional programming model is challenging. Therefore, Epiphany has never gained enough attention in the mainstream general-purpose processor market.

In 2011, Adapteva, a kick-starter company, introduced the first processor based on the Epiphany architecture (Figure 5.2). It contains a 16 RISC core Epiphany chip, expandable to be used in a 256 multi-chip platform (4096 cores in total). The chip is implemented in a 65nm technology node and consumes less than 2 Watts. A few months later, Adapteva introduced a bigger version of the processor with 64 cores. The latest version of the processor [2] was announced in 2016 and contains 1024 cores.

Despite the failure of Epiphany in the general-purpose compute domain, it has a very similar architecture to the neuromorphic processors which were introduced a few years later (e.g., SpiNNaker in 2013 [3], IBM TrueNorth in 2015 [4], Intel Loihi in 2018 [5], BrainChip AKIDA in 2019 [6] and GML NeuronFlow in 2020 [7]). The main goal of neuromorphic engineering is to build a brain-inspired processor to execute variations of Spiking Neural

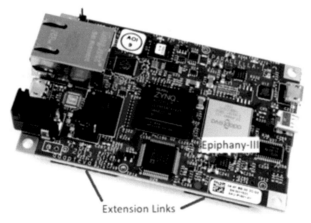

Figure 2.2 Adapteva launched an $99 Epiphany-III based single board computer as their first product.

Network (SNN) algorithms for real-time sensory signal processing. Programming to implement neural networks using conventional programming models and compilers is difficult (and inefficient), which resulted in a new paradigm shift in the programming models. A neural network usually contains neurons (as the processing unit) and weighted synapses/axons to connect the neurons in a graph like architecture. Therefore, several new graph-based programming models (like TensorFlow from Google and PyTorch from Facebook) are introduced to efficiently execute such applications.

The architecture is made up of eNode processing cores and eMesh routers to build connectivity networks. Each eNode contains a RISC processor (1GHz, with an integer and a floating-point ALU and a 64-word register file), 4 memory banks (each $64b \times 1024w$) to store data (like synaptic weights and neuron states), and the instructions (like the neuron model) locally, a Network Interface (NI), Direct Memory Access (DMA) to handle incoming/outgoing packets, a few general timers (for example to implement periodic leakage) and a memory BUS interconnect which allows access to each memory bank simultaneously. The eMesh routers handle 3 separate networks. A high-performance network for sending one packet of data (spike) to the other cores with the maximum speed of one packet per clock cycle) and two lower performance networks (one for reading from another core's memory and one for off-chip communication) are introduced to make the programming easier. These programming models allow for easy splitting of the computational load over several processing units and mapping synaptic connectivity into the NoC. Therefore, they are a good fit for architectures like Epiphany.

Like the other neuromorphic architectures, Epiphany is extremely scalable, performs near memory processing, is optimized for local data movement (local connectivity) and asynchronous processing. The eMesh network is flexible enough to time multiplex any arbitrary synaptic connections. Besides, the eCores are flexible enough to implement different neuron models. Most importantly, the architecture is straightforward, which allows easy design space exploration and benchmarking. Finally, unlike all the other neuromorphic platforms it is accessible and affordable which makes it a suitable platform for benchmarking new neuromorphic platforms and innovative ideas.

2.2 Comparison with a Few Well-Known Digital Neuromorphic Platforms

Probably the SpiNNaker architecture [3] (introduced in 2013) is the most similar neuromorphic platform to Epiphany. SpiNNaker contains several ARM cores as the processing units connected through an advanced asynchronous packet-switched network.

Therefore, like Epiphany, the processing core is very flexible and can implement different neuron models with various mapping schemes. Unlike Epiphany, each SpiNNaker chip contains only one router, with a higher complexity level than the Epiphany's eMesh router. The SpiNNaker's NoC allows for multi-casting (using source-based addressing with a programmable routing table), which is an optimization on top of the plain mesh NoC.

Contrary to SpiNNaker, IBM TrueNorth [4] (introduced in 2015) uses a plain mesh packet-switched network but with optimized processing cores. Therefore, the NoC in IBM TrueNorth is very similar to the Epiphany. Each core in the TrueNorth architecture is fixed to emulate 256 neurons, and each neuron with 256 input synapse (a crossbar architecture) and a single output axon (connectable to 256 neurons in any other core). The cores update all the neurons every 1*ms*. The synaptic weights are limited to be binary. This optimized processing core resulted in an ultra-low-power neuron update (about 26pJ). However, having such constrained cores makes the deployment of many neuromorphic applications either impossible or inefficient.

In Intel Loihi [5] (introduced in 2018), the processing cores are more flexible than TrueNorth, and the interconnect is a simple packet-switched mesh. Each core in Loihi emulates 1024 neurons with a fixed neuron model, but the number of input synapses to each neuron and their resolution is flexible (1kb of synaptic memory per neuron). The number of output axons is also flexible, and one axon can be shared among many neurons. Loihi cores accelerate a bio-inspired learning algorithm. The cost of these flexibility is having a higher neuron update energy (about 80pJ) in comparison with the TrueNorth (while using a better technology node).

In addition to the three previous research platforms, many companies started to build neuromorphic processors for commercial purposes. For example, BrainChip AKIDA (introduced in 2019) and GML NeuronFlow (introduced in 2020) have similar architectures to Loihi.

One of the features in the research of neuromorphic chips is asynchronous processing and communication. In Loihi, the asynchronization level is inside the core's logic blocks. In SpiNNaker and TrueNorth, the cores are working asynchronously with each other in a Globally Asynchronous Locally Synchronous (GALS) structure. In Epiphany, NeuronFlow, and AKIDA, the asynchronousity level is pushed toward the boundaries of the chip (asynchronous chip to chip connectivity). Despite where is the boundary of asynchronousity, it is essential for scalability.

Nevertheless, in all the mentioned architectures, the cores still work individually with each other. Therefore, the implementation of a globally synchronous algorithm is not optimal in neuromorphic architectures.

2.3 Major Challenges in Neuromorphic Architectures

Since neuromorphic architecture design aims to follow the principles of bio-inspired processing mechanism in the available nano-electronic technologies, facing several challenges that result from the platform constraints is normal. Many innovative schemes have been introduced to overcome the difficulties of developing neuromorphic technology and spiking neural network algorithm design. These challenges are discussed below.

2.3.1 Memory Allocation

One of the main challenges in neuromorphic design is the available amount of local memory near or inside the processing element where the data is consumed. In the brain, there is no separation between memory and computation. This feature eliminates a) the memory bandwidth bottleneck issue and b) the high cost of data movement between the processing and a far-away memory block. To mimic this feature, neuromorphic chips use distributed memories near or inside the processing elements (to keep the synaptic weights and neuron states close to the processing unit). However, the onchip memory made by using the conventional SRAM memory technology is not area-efficient (compared to DRAM and Flash) and therefore expensive. Besides using a new denser memory technology [8], one of the solutions to overcome this problem is the proper memory management and maximum reuse of the memory bits.

The important elements to be stored in each processing core are the spike queue(s), synaptic weights, neuron states, and axons (destination addresses). The depth and width of these memories heavily depend on the executable neuron model and supported connectivities. Table 5.1 shows the memory allocations in different neuromorphic chips.

Flexibility in the memory allocations allows for optimized mapping of a neural network in the processor. Different neurons in the neural network have a different number of inputs/outputs and different amounts of activities. Some neuromorphic chips allow flexible parameter resolution to trade-off accuracy and SNN size [5]. Since the range of the parameters is sometimes more important than the resolutions of the parameters, using smaller floating-point representations (like BrainFloat16 [9]) may results in better accuracy

Table 2.1 Memory fragmentations in some digital large-scale neuromorphic chips

Architecture	Total Memory	Spike queue	Neurons	Synapses	Axons
TrueNorth [4]	110kb fixed allocations (426b per neuron)	256*16b	256 fixed neuron type	256*256*1b	256*26b (1 per neuron)
Loihi [5]	2Mb	N/A	1024 fixed neuron type	1Mb Flexible resolution (1b to 9b) Weight sharing	4k flexibly shared
NeuronFlow [7]	120kb	N/A	1024 few neuron types	1k*8b Weight sharing	1k flexibly shared
SpiNNaker [3]	768kb 256kb instruction memory 512kb data memory	Flexible	Flexible pro-grammable neuron type	Flexible resolution only integer ALU	Flexible
Epiphany [1]	256kb in 4 banks, each with 64b data width	Flexible	Flexible pro-grammable neuron type	Flexible resolution Int/float ALUs	Flexible

and power/area performance than using a larger inter (like int32) format. Therefore, it is possible to trade-off the memory footprint and complexity of the operations.

Another method to use the memory space efficiently is to store a compressed form of the parameters when there is a high amount of sparsity in the synaptic weight tensor [10]. Weight sharing is another method to efficiently use the memory for spiking Convolutional Neural Networks (sCNN) [5] [7].

The Epiphany contains 256kb of memory per core and is the most flexible architecture in Table 5.1. In the table N/A means we could not find the data publicly. Axons are the destination core addresses to route spikes from a neuron. All the numbers in this table are for a single processing core inside the mentioned neuromorphic chip. All the above-mentioned schemes can be implemented in Epiphany to optimally use the memory space. To demonstrate

Table 2.2   Mapping LeNet-5 neural network (with binary weights) in different neuromorphic architectures

Architecture	Number of used neurons	Average number of synapses per neuron	Number of individual stored weights	Number of used cores	Total memory used
LeNet-5 (before deployment)	6518	144.5	61k	-	-
TrueNorth [4]	40k	256	941k (144.5×6518)	155	17Mb
Loihi [5]	6518	1024	61k	7	14Mb
NeuronFlow [7]	6518	1024	61k	7	840kb
SpiNNaker [3]	6518	144.5	61k	1	768kb
Epiphany	6518	144.5	61k	1	256kb

the value of flexibility for efficient use of memory, in Table 5.2 we show the result of mapping the binary LeNet-5 [11] into the above-mentioned neuromorphic architecture. The average pooling layers are optimized out in the mapping (as average pooling is a linear operation and does not consume stateful neurons). The mappings are hand optimized with only memory constraint. In TrueNorth, several neurons need to be combined to make a single neuron with enough synapses and axons. Also, since weight-sharing is not used, the weight for each synapse needs to be stored individually. In the flexible architectures, the neuron states are assumed to be 16b, without refractory mechanism and with a single threshold per channel. Mapping in SpiNNaker is done with the "Convnet Optimized Implementation" which is described in [12]. Total memory used is (*number of cores × memory per core*).

2.3.2 Efficient Communication

Using a packet to communicate spikes between cores can be very inefficient. A spike packet that carries a single bit of data (spike) contains several bits for the address. For example, a spike packet in SpiNNaker contains 44b of data to communicate a single binary spike in the AER format [13]. There are several possible solutions to reduce the number of bits for communicating spikes. One solution is to use a more complex neuron model (for example [14] and [15]) with a lower firing rate (trading off operation complexity with the number of packets). Another solution is to compress several spikes into one event. For instance, when the destination core for several packets is the same, we can compress them easily in one hyper-packet. Epiphany's packets are fixed in size (104b packet with a 64b payload data) but the format of payload data is programmable.

TrueNorth [4] and NeuronFlow [7] use a relative addressing scheme which allows reducing the number of bits for the destination address in the packet when a limited communication range is acceptable. For example, in a platform with 4096 cores, if the destination address contains only 4b, a core can only communicate with 16 neighbouring cores which might be sufficient for many applications. This results in a saving of 8b per packet (from 12b address in a 4096-cores system to 4b-address). Another method to reduce the number of packets is the multi-casting feature which is used in SpiNNaker [3]. In this case, a core can only send one spike out and this spike will be multicasted in the NoC and near the destination cores. Epiphany uses the basic mesh NoC interconnect which is a shortcoming but contributes to its simplicity.

2.3.3 Mapping SNN onto Hardware

An optimized mapping algorithm can reduce the memory footprint (by performing maximum sharing of parameters), balance the loads in different cores (as not all the neurons in an SNN are equally active) and reduce the core-to-core communications (since it is expensive in terms of power consumption and latency). Having a flexible number of neurons per core and synapses per neuron allows the mapping optimizer to find a better solution. The Epiphany platform can be used to benchmark different mapping algorithms in the neuromorphic domain because of its flexible and unified memory architecture.

2.3.4 On-chip Learning

On-chip learning is supported as a futuristic feature in some neuromorphic chips (like Loihi [5] and AKIDA [16]). However, implementation a hardware acceleration for on-chip learning is challenging. First, because the algorithm domain is very dynamic (experimental), it is difficult to find a suitable algorithm for a wide range of applications. Second, many applications can be pre-trained and only require fine-tuning after deployment. Therefore, the learning acceleration might be used only for a few last layers of the neural network (after general feature extraction layers). Epiphany does not have a hardware accelerated learning engine, but it allows for software implementation of those algorithms and therefore benchmarking the new learning algorithms.

2.3.5 Idle Power Consumption

One of the challenges in event-based neuromorphic processors is the power consumption when the cores are in the idle state (no event to be processed). It is reported that around 30% of power consumption for TrueNorth [4] and Loihi [17] is the idle power. This can be even worse when the application is sparser. It is possible to reduce idle power consumption by using asynchronous design or clock gating when no input spike is processed. Also, using a non-volatile memory technology helps to reduce leakage in the memory cells (since neuromorphic chips are mostly memory dominant). Epiphany supports dynamic clock gating for the processing cores. In this case, a core can only wake up by an interrupt (for example, receiving a new input packet).

2.4 Measurements from Epiphany

In this section, we present some of our measurements using the Epiphany processor to provide a sense of its performance for possible neuromorphic applications.

We implemented a neural network with a Leaky integrate and Fire (LIF) neuron model with different parameters in Epiphany and measured processing time for different processes, which can be used as a reference for assessment of Epiphany when one wants to use it as a neuromorphic processor. Our measurements in this work consider the processing time (no power measurement) and are performed using the hardware timers inside the cores.

The compiled instructions (not hand-optimized) for our experiments took around 52kb of the used cores' memory. Since the instruction code is almost similar for all the cores, it will be copied in each core's memory. It is therefore recommended to use bigger cores (more memory), so instruction memory takes only a small fraction of the total memory and is used for a higher number of neurons.

Figure 5.3 shows a flowchart of our neuron model with the processing cycle time attached to each block, where N is the number of neurons, F is the number of firings, X is the neuron state, W is the synaptic weight, Thr is the firing threshold, Time is the current time (read from Timer), LFT is the last firing time (stored per neuron), Ref is the refractory time and LR is the leak rate.

An input spike enters the eCore through the DMA and interrupts the RISC core. Then a process handles this spike and puts it in a FIFO (made with

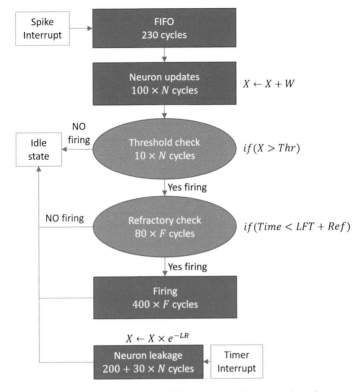

Figure 2.3 Flow chart of processing a LIF neuron with processing time measured in Epiphany.

a software process). Thereafter, the target neurons will get updated. After updates, the threshold of neurons is checked, and the refractory check is executed for each firing. If both checks pass, the firing process starts, and the RISC core commands to the DMA to transmit a spike packet. Membrane leakage is also an independent process that starts with a timer interrupt.

Each cycle takes 1ns when using a 1GHz clock frequency. For example, processing a single spike from the first convolutional layer of LeNet-5 to the second convolutional layer requires to update $16 \times 5 \times 5$ neurons. When the second layer is implemented in a core and 1% of the updated neurons fire, the processing time takes around 46us. The leak process on all these 400 neurons takes around 12us. Our measurements are averaged over many experiments and therefore the numbers in this figure are reasonable estimations. Since the neuron model is programmable, one may decide to remove some of the components (like refractory) or make it more complex (for example by introduction of an individual threshold for every neuron)

In Figure 5.3 we showed that updating a neuron with a single spike takes around 120ns on average. We know that TrueNorth can update all of the neurons in a core every 1ms, to be suitable for real-time neuromorphic applications. If we assume a reasonable sparsity in the input spikes in each time-step (32 input spikes per neuron with 256 input synapses), with 120ns update time, Epiphany can also process the 256 neurons in less than 1ms.

2.5 Conclusion

This article demonstrates that the Epiphany processor is compatible with neuromorphic computing. Overall, it has a similar architecture to the well-known neuromorphic processors and is flexible enough for the implementation of new ideas. Unlike Epiphany, all the mentioned neuromorphic processors contain optimized elements that add complexity to the architecture and make it less flexible to be a reference benchmarking architecture (flexibility vs efficiency trade-off). For example, having a fixed number of neurons per core (in TrueNorth, Loihi, and NeuronFlow) does not allow for optimized resource management during mapping. Also, having an accelerated learning mechanism (in Loihi) may be unnecessary for many applications. Additionally, suppose one wants to know the performance improvement of the SpiNNaker processor due to its optimized NoC. In that case, Epiphany is an excellent platform to compare to, due to its simplicity and flexibility.

As mentioned, not having any accelerator makes the epiphany less efficient compared to the accelerated architectures (like Loihi), but it increases its value for benchmarking the performance improvement of any accelerators.

We have implemented a neural network system and measured the processing time for different components of the LIF neuron model. It is already visible that some small improvements (like having a hardware FIFO) can improve the performance of the system. Increasing the size of the core results in better memory saving, but the designer should scale the performance of the cores as well (by the implementation of the schemes like multi-threading [5] and SIMD, as it is implemented in the forthcoming SpiNNaker2.0 platform [18]). Other improvements (like adding a more suitable interconnect) can be examined and is a topic for our future research. All source code used to benchmark the system and perform hands-on experiments is freely available upon request ({amirreza.yousefzadeh, gert-jan.vanschaik}@imec.nl)

Acknowledgements

This technology is partially funded and initiated by the Netherlands and European Union's Horizon 2020 research and innovation projects **TEMPO** (ECSEL Joint Undertaking under grant agreement No 826655) and **ANDANTE** (ECSEL Joint Undertaking under grant agreement No 876925).

References

[1] A. Olofsson, et al., Kickstarting high-performance energy-efficient manycore architectures with epiphany, in 2014 48th Asilomar Conference on Signals, Systems and Computers, IEEE, 2014, pp. 1719–1726.

[2] A. Olofsson, Epiphany-v: A 1024 processor 64-bit risc system-on-chip, arXiv preprint arXiv:1610.01832.

[3] E. Painkras, et al., Spinnaker: A 1-w 18-core system-on-chip for massively-parallel neural network simulation, IEEE Journal of Solid-State Circuits 48 (8) (2013) 1943–1953.

[4] F. Akopyan, et al., Truenorth: Design and tool flow of a 65 mw 1 million neuron programmable neurosynaptic chip, IEEE transactions on computer-aided design of integrated circuits and systems 34 (10) (2015) 1537–1557.

[5] M. Davies, et al., Loihi: A neuromorphic manycore processor with on-chip learning, IEEE Micro 38 (1) (2018) 82–99.

[6] M. Demler, Brainchip akida is a fast learner, spiking-neural-network processor identifies patterns in unlabeled data, Microprocessor Report (2019).

[7] O. Moreira, et al., Neuronflow: a neuromorphic processor architecture for live ai applications, in 2020 Design, Automation & Test in Europe Conference & Exhibition (DATE), IEEE, 2020, pp. 840–845.

[8] E. Miranda, J. Suñé, Memristors for neuromorphic circuits and artificial intelligence applications (2020).

[9] N. P. Jouppi, et al., In-datacenter performance analysis of a tensor processing unit, in: Proceedings of the 44th Annual International Symposium on Computer Architecture, 2017, pp. 1–12.

[10] V. Sze, Y.-H. Chen, T.-J. Yang, J. S. Emer, Efficient processing of deep neural networks, Synthesis Lectures on Computer Architecture 15 (2) (2020) 1–341.

[11] Y. LeCun, et al., Lenet-5, convolutional neural networks, URL: http://yann. lecun. com/exdb/lenet 20 (5) (2015) 14.

[12] A. Yousefzadeh, et al., Performance comparison of time-step-driven versus event-driven neural state update approaches in spinnaker, in 2018 IEEE International Symposium on Circuits and Systems (ISCAS), IEEE, 2018, pp. 1–4.

[13] A. Yousefzadeh, et al., Fast predictive handshaking in synchronous FPGAs for fully asynchronous multisymbol chip links: Application to spinnaker 2-of-7 links, IEEE Transactions on Circuits and Systems II: Express Briefs 63 (8) (2016) 763–767.

[14] A. Yousefzadeh, et al., Asynchronous spiking neurons, the natural key to exploit temporal sparsity, IEEE Journal on Emerging and Selected Topics in Circuits and Systems 9 (4) (2019) 668–678. doi:10.1109/JETCAS.2019.2951121.

[15] B. Yin, et al., Effective and efficient computation with multiple-timescale spiking recurrent neural networks, in International Conference on Neuromorphic Systems 2020, ICONS 2020, Association for Computing Machinery, New York, NY, USA, 2020. doi:10.1145/3407197.3407225.

[16] S. Thorpe, et al., Method, digital electronic circuit, and system for unsupervised detection of repeating patterns in a series of events, US Patent App. 16/349,248 (Sep. 19, 2019).

[17] P. Blouw, et al., Benchmarking keyword spotting efficiency on neuromorphic hardware, in: Proceedings of the 7th Annual Neuro-inspired Computational Elements Workshop, 2019, pp. 1–8.

[18] C. Mayr, S. Höppner, and S. Furber (2019). SpiNNaker 2: a 10 million core processor system for brain simulation and machine learning-keynote presentation. In Communicating Process Architectures 2017 & 2018 277-280, IOS Press, 2019.

3

Temporal Delta Layer: Exploiting Temporal Sparsity in Deep Neural Networks for Time-Series Data

Preetha Vijayan[1,2], Amirreza Yousefzadeh[2], Manolis Sifalakis[2], and Rene van Leuken[1]

[1]TU Delft, Netherlands
[2]imec, Netherlands

Abstract

Real-time video processing using state-of-the-art deep neural networks (DNN) has managed to achieve human-like accuracy in the recent past but at the cost of considerable energy consumption, rendering them infeasible for deployment on edge devices. The energy consumed by running DNNs on hardware accelerators is dominated by the number of memory read/writes and multiply-accumulate (MAC) operations required. This work explores the role of activation sparsity in efficient DNN inference as a potential solution. As matrix-vector multiplication of weights with activations is the most predominant operation in DNNs, skipping operations and memory fetches where (at least) one of them is a zero can make inference more energy efficient. Although spatial sparsification of activations is researched extensively, introducing and exploiting temporal sparsity has received far less attention in DNN literature. This work introduces a new DNN layer (called temporal delta layer) whose primary objective is to induce temporal activation sparsity during training. The temporal delta layer promotes activation sparsity by performing delta operation that is aided by activation quantization and l_1 norm based penalty to the cost function. As a result, the final model behaves like a conventional quantized DNN with high temporal activation sparsity during inference. The new layer was incorporated into the standard ResNet50 architecture to be trained and tested on the popular human action recognition

DOI: 10.1201/9781003377382-3

dataset, UCF101. The method resulted in a 2x improvement in activation sparsity, with a 5% reduction in accuracy.

3.1 Introduction

DNNs have lately managed to successfully analyze video data to perform action recognition [1], object tracking [2], object detection [3], etc., with human-like accuracy and robustness. Unfortunately, DNNs' high accuracy comes with considerable costs, in terms of computation and memory consumption, resulting in high energy consumption. This makes them unsuitable for always-on edge devices.

Techniques such as network pruning, quantization, regularization, and knowledge distillation [4] [5] have helped reduce model size over time, resulting in less compute and memory consumption overall. Sparsity is a prominent aspect in all of the aforementioned methods. This is significant because sparse tensors allow computations involving zero multiplication to be skipped. They are also easy to store and retrieve in memory. In the DNN literature, structural sparsity (of weights) and spatial sparsity (of activations) are well-studied topics [6]. However, while being a popular concept in neuromorphic computing, temporal activation sparsity has received less attention in the context of DNN.

This work applies the concept of change or delta based processing to the training and inference phases of deep neural networks, drawing inspiration from the human retina [7]. DNN inference, which processes each frame independently with no regard to the temporal correlation is dense and obscenely wasteful. Whereas, processing only the changes in the network can lead to zero-skipping in sparse tensor operations minimizing the redundant operations and memory accesses.

Therefore, the proposed methodology in this work induces temporal sparsity to theoretically any DNN by incorporating a new layer (named temporal delta layer), which can be introduced in a DNN at any phase (training, refinement, or inference only). This new layer can be integrated to an existing architecture by positioning it after all or some of the ReLU activation layers as deemed beneficial (see Figure 3.1). The inclusion of this layer does not necessitate any changes to the preceding or following layers. Furthermore, the new layer adds a novel sparsity penalty to the overall cost function of the DNN during the training phase. This l_1 norm based penalty minimizes the activation density of the delta maps (i.e., temporal difference between two consecutive feature maps). Apart from that, the new layer is compared

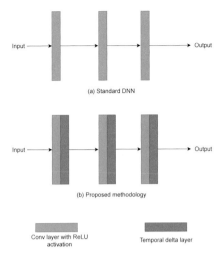

Figure 3.1 (a) Standard DNN, and (b) DNN with proposed temporal delta layer

in conjunction with two activation quantization methods, namely fixed-point quantization (FXP) and learned step-size quantization (LSQ).

3.2 Related Works

Although DNNs are in essence bio-inspired, they have not been able to find the balance between power consumption and accuracy yet, especially while dealing with computationally heavy streaming signals. On the other hand, the brain's neocortex handles complex tasks like sensory perception, planning, attention, and motor control while consuming less than 20 W [8]. Scalable architecture, in-memory computation, parallel processing, communication using spikes, low precision computation, sparse distributed representation, asynchronous execution, and fault tolerance are some of the characteristics of the biological neural networks that can be leveraged to bridge the energy consumption gap between the brain and DNNs [9]. Among these, the proposed methodology focuses on the viability of using sparsity within DNNs to achieve energy efficiency. During a matrix-vector multiplication between a weight matrix and an activation vector, zero elements in the tensor can be skipped leading to computational as well as memory access reduction (see Figure 1.2).

There are broadly two types of sparsity available in DNNs: weight sparsity (related to the interconnect between neurons) and activation sparsity

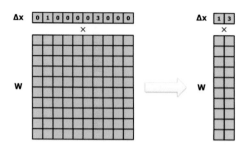

Figure 3.2   Sparsity in activation (Δx) drastically reduce the memory fetches and multiplications between Δx and columns of weight matrix, W, that correspond to zero [10].

(related to the number of neurons). Furthermore, activation sparsity can be categorised into spatial and temporal sparsity, which exploits the spatial and temporal correlation within the activations, respectively, [11]. Unlike weight and spatial sparsity [12, 13, 14, 15], exploiting the temporal redundancy of DNNs while processing streaming data as a means to reduce energy consumption is a relatively less explored idea. Exploiting temporal sparsity translates to skipping re-calculation of a function when its input remains unchanged since the last update.

One of the methods to exploit temporal sparsity is to use the compressed representation (like H.264, MPEG-4, etc.) of videos at the input stage itself. These compression techniques only retain a few key-frames completely and reconstruct others using motion vectors and residual error, thus using temporal redundancy [16] [17]. Another path includes finding a neuron model which is somewhere in between "frame-based DNN" and "event-based spiking neural networks". This work is an attempt in the aforementioned direction. A similar work, CBInfer [18], proposes replacing all spatial convolution layers in a network with change-based temporal convolution layers (or CBconv layers). In this, a signal change is propagated forward only when a certain threshold is exceeded. Likewise, [19] tapped into temporal sparsity by introducing Sigma-Delta Networks, where neurons in one layer communicated with neurons in the next layer through discretized delta activations. An issue when it comes to CBInfer is the potential error accumulation over time as the method is threshold-based. If the neuron states are not reset periodically, this threshold can cause drift in the approximation of the activation signal and degrade the accuracy. Whereas, sigma-delta scheme experiments on smaller datasets like temporal MNIST, which might not be a reliable confirmation of the method's effectiveness.

3.3 Methodology

In video-based applications, traditional deep neural networks rely on frame-based processing. That is, each frame is processed entirely through all the layers of the model. However, there is very little change in going from one frame to the next through time, which is called temporal locality. Therefore, it is wasteful to perform computations to extract the features of the non-changing parts of the individual frame. Taking that concept deeper into the network, if feature maps of two consecutive frames are inspected after every activation layer throughout the model, this temporal overlap can be observed. Therefore, this work postulates that temporal sparsity can be significantly increased by focusing the inference of the model only on the changing pixels of the feature maps (or deltas).

3.3.1 Delta Inference

This work introduces a new layer that calculates the delta (or difference) between two temporally consecutive feature maps and quantifies the degree of these changes at only relevant locations in the frame. Since zero changes are not propagated through the layer, the role of this layer may be perceived as "analog event propagation". It is considered an "analog event" as it is not the presence of change, but the magnitude of change that is propagated through.

To better understand it mathematically, in a standard DNN layer, the output activation is related to its weights and input vector through Eq. 3.1 and 3.2.

$$Y_t = W X_t + B \qquad (3.1)$$

$$Z_t = \sigma(Y_t) \qquad (3.2)$$

where W and B represent the weights and bias parameters, Xt represents the input vector, and Yt represents the transitional state. Then, Zt is the output vector which is the result of $\sigma(.)$ - a non-linear activation function. t indicates that the tensor has a temporal dimension. However, in the temporal delta layer, weight-input multiplication transforms into,

$$\Delta Y_t = W \Delta X_t = W(X_t - X_{t-1}) \qquad (3.3)$$

$$Y_t = \Delta Y_t + Y_{t-1}$$
$$= W(X_t - X_{t-1}) + W(X_{t-1} - X_{t-2}) + \ldots + Y_0, \quad where \ Y_0 = B$$
$$= W X_t + B,$$

$$(3.4)$$

$$\Delta Z_t = Z_t - Z_{t-1} = \sigma(Y_t) - \sigma(Y_{t-1}), \quad where \ \sigma(Y_0) = 0 \qquad (3.5)$$

In Eq. 3.3, instead of using X_t directly, only changes or ΔX_t are multiplied with W. Using the resulting ΔY_t, the corresponding Y_t can be recursively calculated with Eq. 3.4, where Y_{t-1} is the transitional state obtained from the previous calculation. Eq. 3.5 is the final delta activation output that is passed onto the next layer.

Another notable difference between the standard DNN layer and the proposed layer is the role of bias. In delta based inference, bias is only used as an initialization for the transitional state, Y_0 in Eq. 3.4. However, since bias tensors do not change over time, their temporal difference is zero and is removed from Eq. 3.3.

Now, as the input video is considered temporally correlated, the expectation is that ΔX_t and by association ΔZ_t are also temporally sparse. In essence, the temporal sparsity between consecutive feature maps is cast on the spatial sparsity of the delta map that is propagated. Additionally, Y_t in Eq. 3.1 and 3.4 are always equal. This indicates that as long as the input is the same, both standard DNN and temporal delta layer based DNN provide the same result at any time step.

3.3.2 Sparsity Induction Using Activation Quantization

As shown in Figure 3.3, there is temporal redundancy evident in feature maps of two consecutive frames. However, if looked closely, it can be observed that these feature maps are similar but not identical as shown in Figure 3.3a and 3.3b. Therefore, if two such consecutive feature maps are subtracted, the resulting delta map has many near zero values, thus restricting the potential increase in temporal sparsity (Figure 3.3c). This is mainly due to the higher precision available in the floating point representation (FP32) of the activations. For example, in IEEE 754 representation, a single-precision 32-bit floating point number has 1 bit for sign, 8 bits for the exponent and 23 bits for the significant. It, not only, leads to a very high dynamic range, but also, increases the resolution or precision for numbers close to 0. The number nearest to 0 is about $\pm 1.4 \times 10^{-45}$. Therefore, due to high resolution, two similar floating point values have difficulty going to absolute zero when

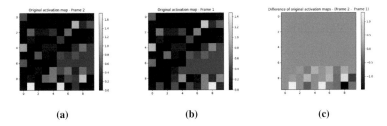

Figure 3.3 Demonstration of two temporally consecutive activation maps leading to near zero values (rather than absolute zeroes) after delta operation.

subtracted. A plausible solution to decrease the precision of the activations is to use quantization.

In this work, a post-training quantization method (fixed point quantization [20]) and a quantization aware training method (learnable step size quantization [21]) are considered for comparison as a temporal sparsity facilitator for the new layer.

3.3.2.1 Fixed Point Quantization

In this method, the floating point numbers are quantized to integer or fixed point representation [20]. Unlike floating point, in fixed point representation, the integer and the fractional part have fixed length. This limits both range and precision. That is, if more bits are used to represent the integer part, it subsequently decreases the precision and vice versa.

Method:
Firstly, a bitwidth is defined to which the 32-bit floating parameter is to be quantized, BW. Then, the number of bits required to represent the unsigned integer part of the parameter (x) is calculated as shown in Eq. 3.6.

$$I = 1 + \lfloor log_2 \left(\max_{1 < i < N} |x| \right) \rfloor \tag{3.6}$$

A positive value of I means that I bits are required to represent the absolute value of the integer part, while a negative value of I means that the fractional part has I leading unused bits. Now, it is known that 1 bit is for sign, so the number of fractional bits, F, is given by Eq. 3.7.

$$F = BW - I - 1 \tag{3.7}$$

Considering the parameters, BW - bitwidth, F - fractional bits, I - integer bits, and S - sign bit, Eq. 3.8 maps the floating point parameter x to the fixed point by,

$$Q(x) = \frac{C(R(x.2^F), -t, t)}{2^F} \tag{3.8}$$

where $R(.)$ is the round function, $C(x, a, b)$ is the clipping function, and t is defined as,

$$t = \begin{cases} 2^{BW-S}, & BW > 1 \\ 0 & BW \leq 1 \end{cases}$$

Possible Drawback of Fixed Point Quantization:
Fixed point quantization, as shown above, is a fairly straightforward mapping scheme and is easy to be included in the model training process during the forward pass before the actual delta calculation. However, it poses a limitation to the extent of quantization possible without sacrificing accuracy. Typically, an 8-bit quantization can sustain floating point accuracy with this method, but if the bitwidth goes below 8 bits, the accuracy starts to deteriorate significantly. This is because, unlike weights, activations are dynamic and activation patterns change from input to input making them more sensitive to harsh quantization [22]. Also, quantizing the layers of a network to the same bitwidth can mean that the inter-channel behaviour of the feature maps are not captured properly. Since the number of fractional bits is usually selected depending on the maximum activation value in a layer, this type of quantization tends to cause excessive information loss in channels with a smaller range.

3.3.2.2 Learned Step-Size Quantization

Quantization aware training is the most logical solution to the aforementioned drawback as it can potentially recover the accuracy in low bit tasks given enough time to train. Therefore, a symmetric uniform quantization scheme is considered called Learned Step size Quantization (LSQ). This method considers the quantizer itself as a trainable parameter which is trying to minimize the task loss using backpropagation and stochastic gradient descent. This serves two purposes: (a) step size, which is the width of quantization bins, gets to be adaptive through the training according to the activation distribution. It is vital to find an optimum step size because, as shown in Figure 3.4, if the step size is too small or too large, it can lead to the quantized

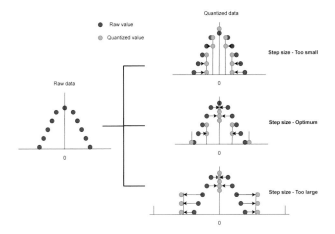

Figure 3.4 Importance of step size in quantization: on the right side, in all three cases, the data is quantized to five bins with different uniform step sizes. However, without optimum step size value, the quantization can detrimentally alter the range and resolution of the original data.

data being a poor representation of the raw data. (b) as the step size is a model parameter, it is also directly seeking to improve the metric of interest, i.e. accuracy.

Method:

Given: x - the parameter to be quantized, s - step size, Q_N and Q_P - number of negative and positive quantization levels respectively, and q(x;s) is the quantized representation with the same scale as x,

$$q(x; s) = \begin{cases} \left\lfloor \frac{x}{s} \right\rceil .s, & \text{if } -Q_N \leq \frac{x}{s} \leq Q_P \\ -Q_N.s, & \text{if } \frac{x}{s} \leq -Q_N \\ Q_P.s, & \text{if } \frac{x}{s} \geq Q_P \end{cases} \qquad (3.9)$$

where $\lfloor a \rceil$ rounds the value to the nearest integer. Considering the number of bits, b, to which the data is to be quantized, $Q_N = 0$ for unsigned and $Q_N = 2^{b-1}$ for signed data. Similarly, $Q_P = 2^{b-1}$ for unsigned and $2^{b-1} - 1$ for signed data.

Modified LSQ:

In this work, the original LSQ method is slightly modified to remove the clipping function from the equations as (a) the bitwidth, b, required to calculate

Q_N and Q_P is not known. This is because the bitwidth is not pre-defined and is determined using the activation statistics of each layer while training which leads to a mixed precision model, which is more advantageous, and (b) clipping leads to accuracy drop as it alters the range of the activation. That is, if activations are clipped during training, there could be a significant difference between the real-valued activation value and the quantized activation value, which in turn affects the gradient calculations and, therefore the SGD optimization.

Thus, in temporal delta layer, the forward pass of the quantization includes only scaling, rounding and de-scaling and can be mathematically expressed as,

$$q(x; s) = \lfloor \frac{x}{s} \rceil . s \tag{3.10}$$

The gradient of the Eq. 3.10 for backpropagation is given by Eq. 3.11.

$$\nabla_s q(x; s) = \lfloor \frac{x}{s} \rceil - \frac{x}{s} \tag{3.11}$$

3.3.3 Sparsity Penalty

Quantized delta map, created using the above-mentioned methods, in itself has a fair number of absolute zeroes (or sparsity) available. However, like the biological brain, learning can help in increasing this sparsity further. The inspiration for this came from an elegant set of experiments performed by Y. Yu et al. [23]. The experiment showed a particular 30 seconds video to rodent specimens and tracked their activation density during each presentation. It was found that activation density decreased as the number of trials increased, i.e as the learning increased, the active neurons required for inference decreases.

Adapting the said concept to this work, a l_1 norm based constraint is introduced to the loss function. This is termed as the *sparsity penalty*. Therefore, the new cost function can be mathematically expressed as *cost function = task loss + sparsity penalty*, i.e,

$$Cost\ function = Task\ loss + \lambda\ (\frac{l_1\ norm\ of\ active\ neurons\ in\ delta\ map}{total\ number\ of\ neurons\ in\ delta\ map}) \tag{3.12}$$

where task loss minimizes the error between the true value and the predicted value and, sparsity penalty minimizes the overall temporal activation density. The λ mentioned in Eq. 3.12 refers to the penalty co-efficient of the cost function. If λ is too small, the sparsity penalty takes little effect and model accuracy is given more priority and if λ is too large, sparsity becomes the priority leading to very sparse models but with unacceptable accuracy. The key is to find the balance between task loss and sparsity penalty.

3.4 Experiments and Results

In this section, the proposed methodology explained in section 3.4 is analyzed to study how it helps achieve the desired temporal sparsity and accuracy.

3.4.1 Baseline

For baseline, the two-stream architecture [24] was used, with ResNet50 as the feature extractor on both spatial and temporal streams. The dataset used was UCF101, which is a widely used human action recognition dataset of 'in-the-wild' action videos, obtained from YouTube, having 101 action categories [25]. The spatial stream used single-frame RGB images of size (224, 224, 3) as the input, while the temporal stream used stacks of 10 RGB difference frames of size (224, 224, 10 $\times$ 3) as the input. Also, both these inputs were time distributed to apply the same layer to multiple frames simultaneously and produce output that has time as the fourth dimension. Both the streams were initialized with pre-trained ImageNet weights and fine-tuned with an SGD optimizer.

Under the above-mentioned setup, spatial and temporal streams achieved an accuracy of 75% and 70%, respectively. Then, both streams were *average fused* to achieve a final classification accuracy of 82%. Also, in this scenario, both streams were found to have an activation sparsity of $\sim 47\%$.

3.4.2 Experiments

Scenario 1: The setup consecutively places the fixed point based quantization layer and temporal delta layer after every activation layer in the network. The temporal delta layer here also includes a l_1 norm based penalty. The baseline weights were used as a starting point, and all the layers including the temporal delta layer is fine-tuned until acceptable convergence. The hyper-parameters specifically required for this setup were bitwidth (to which the activations were to be quantized) and penalty co-efficient to balance the tussle between task loss and sparsity penalty.

Scenario 2: The setup is similar to the previous scenario except for the activation quantization method used. The previous experiment used fixed precision quantization where all the activation layers in the network were quantized to the same bitwidth. However, this experiment uses learnable step-size quantization (LSQ), which performs channel-wise quantization depending on the activation distribution resulting in mixed-precision quantization of the activation maps.

The layer also introduces a hyperparameter during training (apart from the penalty coefficient mentioned earlier) for the step size initialization. Then, during training, the step size increases or decreases depending on the activation distribution in each channel.

3.4.3 Result Analysis

Table 3.1 and 3.1 show the baseline accuracy and activation sparsity compared against the two scenarios mentioned.

Firstly, when the temporal delta layers with fixed point quantized activations are included in the baseline model, it can be observed that the activation sparsity increases considerably with a slight loss in accuracy in both streams.

Table 3.1   Spatial stream - comparison of accuracy and activation sparsity obtained through the proposed scenarios against the baseline. In the case of fixed point quantization, the reported results are for a bitwidth of 6 bits.

Model setup (Spatial stream)	Accuracy	Activation sparsity
Baseline	75%	48%
Temporal delta layer with fixed point quantization	73%	74%
Temporal delta layer with learned step-size quantization	**69%**	**86%**

Table 3.2   Temporal stream - comparison of accuracy and activation sparsity obtained through the proposed scenarios against the benchmark. In the case of fixed point quantization, the reported results are for a bitwidth of 7 bits.

Model setup (Temporal stream)	Accuracy	Activation sparsity
Baseline	70%	47%
Temporal delta layer with fixed point quantization	68%	67%
Temporal delta layer with learned step-size quantization	**65%**	**89%**

This is because lowering the precision from 32 bits to 8 bits (or less) leads to temporal differences of activations going to absolute zero.

Additionally, the reason for close-to baseline accuracy in the method involving fixed point quantization can be attributed to fractional bit allocation flexibility. That is, as the bitwidth is fixed, the number of integer bits required is decided depending on the activation distribution within the layer, and the rest of the bits are assigned as fractional bits. This makes sure that the precision of the activation is compromised for range. Also, another contributing factor for accuracy sustenance is that the first and the last layers of the model are not quantized, similar to works like [26][27]. This is because the first and last layer has a lot of information density. Those are the layers where input pixels turn into features and features turn into output probabilities, respectively, which makes them more sensitive to quantization.

Although the activation sparsity gain in the case of the temporal delta layer with fixed point quantization is better than the baseline, it is still not sufficiently high as required. In this effort, the bitwidth of the activations are decreased in the expectation of increasing sparsity. However, as the bitwidth goes below a certain value (6 bits for spatial and 7 bits for temporal stream), sparsity increases, but accuracy starts to deteriorate beyond recovery, as shown in Table 3.3. This is because quantizing all layers of a network to the same bitwidth can mean that the inter-channel variations of the feature maps are not fully accounted for. Since the number of fractional bits is usually selected to cover the maximum activation value in a layer, the fixed bitwidth quantization tends to cause excessive information loss in channels with a smaller dynamic range. Therefore, it can be inferred that mixed-precision quantization of activations is a better approach to obtain good sparsity without compromising accuracy.

Table 3.3 Result of decreasing activation bitwidth in fixed point quantization method. For spatial stream, decreasing below 6 bits caused the accuracy to drop considerably. For temporal stream, the same happened below 7 bits.

	Spatial stream		Temporal stream	
Activation bitwidth	**Accuracy (%)**	**Activation sparsity (%)**	**Accuracy (%)**	**Activation sparsity (%)**
32	75	50	70	47
8	75	68	70	65
7	75	71	**68**	**70**
6	**73**	**75**	61	73
5	65	80	-	-

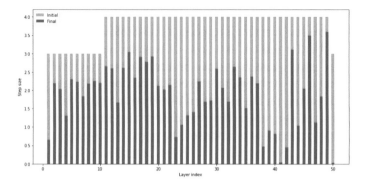

Figure 3.5   Evolution of quantization step size from initialization to convergence in LSQ. As step-size is a learnable parameter, it gets re-adjusted during training to cause minimum information loss in each layer.

Finally, using the temporal delta layer where incoming activations are quantized using learnable step-size quantization (LSQ) gives the best results for both spatial and temporal streams. As the step size is a learnable parameter, it gives the model enough flexibility to result in a mixed precision model, where each channel in a layer has a bitwidth that suits its activation distribution. This kind of channel-wise quantization minimizes the impact of low-precision rounding. It is also evident in Figure 3.5 that as the training nears convergence, the values of the step size differ according to the activation distribution and bitwidth required to represent each layer. Moreover, consistent with the expectation, the first and last layers during training opts for smaller step sizes implying they need more bitwidth for their representation.

Table 3.4   Final results from two-stream network after average fusing the spatial and temporal stream weights. With 5% accuracy loss, the proposed method almost doubles the activation sparsity available in comparison to the baseline.

Model type	Baseline		Proposed method	
	Accuracy (%)	Activation sparsity (%)	Accuracy (%)	Activation sparsity (%)
Spatial stream	75	50	69	86
Temporal stream	70	46	65	89
Two-stream (Average fused)	82	47	**77**	**88**

The weights generated using this method was then average fused to find the final two-stream network accuracy and activation sparsity (Table 3.4). Finally, the proposed method can achieve an overall 88% activation sparsity with 5% accuracy loss.

3.5 Conclusion

Intuitively, the proposed new temporal delta layer projects the temporal activation sparsity between two consecutive feature maps onto the spatial activation sparsity of their delta map. When executing sparse tensor multiplications in hardware, this spatial sparsity can be used to decrease the computations and memory accesses. As shown in Table 3.4, the proposed method resulted in 88% overall activation sparsity with a trade-off of 5% accuracy drop on UCF-101 dataset.

The collateral benefit of the obtained temporal sparsity is that the computations does not increase linearly with the increase in frame rate. In typical DNNs, doubling the frame rate would automatically necessitate doubling the computations. However, in the case of temporal delta layer based model, increasing the frame rate will not only improve the temporal precision of the network but also increase its temporal sparsity limiting the computations required [28].

The downside of using the temporal delta layer is that it requires keeping track of previous activations in order to perform delta operations. As a result, the overall memory footprint grows, putting more reliance on off-chip memory. However, the rising popularity of novel memory technologies (like resistive RAM [29], embedded Flash memory [30], etc.) may improve the cost calculations in the near future.

Disclaimer: This paper is a distillation of the research done by one of the authors as a part of her master thesis and is partially published in chapter 3 of [32]. The complete thesis, along with the results and analysis, is available online [31].

Acknowledgment

This work is partially funded by research and innovation projects TEMPO (ECSEL JU under grant agreement No 826655), ANDANTE (ECSEL JU under grant agreement No 876925) and DAIS (KDT JU under grant agreement No 101007273), SunRISE (EUREKA cluster PENTA2018e-17004-SunRISE) and Comp4Drones (ECSEL JU grant agreement No. 826610).

The JU receives support from the European Union's Horizon 2020 research and innovation programme and Sweden, Spain, Portugal, Belgium, Germany, Slovenia, Czech Republic, Netherlands, Denmark, Norway and Turkey.

References

[1] L. Wang, Y. Xiong, Z. Wang, Y. Qiao, D. Lin, X. Tang, and L. Van Gool, "Temporal segment networks: Towards good practices for deep action recognition," in *European conference on computer vision*, pp. 20–36, Springer, 2016.

[2] K. Chen and W. Tao, "Once for all: a two-flow convolutional neural network for visual tracking," *IEEE Transactions on Circuits and Systems for Video Technology*, vol. 28, no. 12, pp. 3377–3386, 2017.

[3] K. Kang, H. Li, J. Yan, X. Zeng, B. Yang, T. Xiao, C. Zhang, Z. Wang, R. Wang, X. Wang, *et al.*, "T-cnn: Tubelets with convolutional neural networks for object detection from videos," *IEEE Transactions on Circuits and Systems for Video Technology*, vol. 28, no. 10, pp. 2896–2907, 2017.

[4] S. Han, H. Mao, and W. J. Dally, "Deep compression: Compressing deep neural networks with pruning, trained quantization and huffman coding," *arXiv preprint arXiv:1510.00149*, 2015.

[5] G. Hinton, O. Vinyals, and J. Dean, "Distilling the knowledge in a neural network," *arXiv preprint arXiv:1503.02531*, 2015.

[6] W. Wen, C. Wu, Y. Wang, Y. Chen, and H. Li, "Learning structured sparsity in deep neural networks," *arXiv preprint arXiv:1608.03665*, 2016.

[7] M. Mahowald, "The silicon retina," in *An Analog VLSI System for Stereoscopic Vision*, pp. 4–65, Springer, 1994.

[8] J. W. Mink, R. J. Blumenschine, and D. B. Adams, "Ratio of central nervous system to body metabolism in vertebrates: its constancy and functional basis," *American Journal of Physiology-Regulatory, Integrative and Comparative Physiology*, vol. 241, no. 3, pp. R203–R212, 1981.

[9] A. Yousefzadeh, M. A. Khoei, S. Hosseini, P. Holanda, S. Leroux, O. Moreira, J. Tapson, B. Dhoedt, P. Simoens, T. Serrano-Gotarredona, *et al.*, "Asynchronous spiking neurons, the natural key to exploit temporal sparsity," *IEEE Journal on Emerging and Selected Topics in Circuits and Systems*, vol. 9, no. 4, pp. 668–678, 2019.

[10] C. Gao, D. Neil, E. Ceolini, S.-C. Liu, and T. Delbruck, "Deltarnn: A power-efficient recurrent neural network accelerator," in *Proceedings of the 2018 ACM/SIGDA International Symposium on Field-Programmable Gate Arrays*, pp. 21–30, 2018.

[11] O. Moreira, A. Yousefzadeh, F. Chersi, G. Cinserin, R.-J. Zwartenkot, A. Kapoor, P. Qiao, P. Kievits, M. Khoei, L. Rouillard, *et al.*, "Neuronflow: a neuromorphic processor architecture for live ai applications," in *2020 Design, Automation & Test in Europe Conference & Exhibition (DATE)*, pp. 840–845, IEEE, 2020.

[12] J. Frankle and M. Carbin, "The lottery ticket hypothesis: Finding sparse, trainable neural networks," *arXiv preprint arXiv:1803.03635*, 2018.

[13] H. Yang, W. Wen, and H. Li, "Deephoyer: Learning sparser neural network with differentiable scale-invariant sparsity measures," *arXiv preprint arXiv:1908.09979*, 2019.

[14] S. Seto, M. T. Wells, and W. Zhang, "Halo: Learning to prune neural networks with shrinkage," in *Proceedings of the 2021 SIAM International Conference on Data Mining (SDM)*, pp. 558–566, SIAM, 2021.

[15] M. Mahmoud, K. Siu, and A. Moshovos, "Diffy: A déjà vu-free differential deep neural network accelerator," in *2018 51st Annual IEEE/ACM International Symposium on Microarchitecture (MICRO)*, pp. 134–147, IEEE, 2018.

[16] C.-Y. Wu, M. Zaheer, H. Hu, R. Manmatha, A. J. Smola, and P. Krähenbühl, "Compressed video action recognition," in *Proceedings of the IEEE Conference on Computer Vision and Pattern Recognition*, pp. 6026–6035, 2018.

[17] M. Buckler, P. Bedoukian, S. Jayasuriya, and A. Sampson, "Eva^2: Exploiting temporal redundancy in live computer vision," in *2018 ACM/IEEE 45th Annual International Symposium on Computer Architecture (ISCA)*, pp. 533–546, IEEE, 2018.

[18] L. Cavigelli, P. Degen, and L. Benini, "Cbinfer: Change-based inference for convolutional neural networks on video data," in *Proceedings of the 11th International Conference on Distributed Smart Cameras*, pp. 1–8, 2017.

[19] P. O'Connor and M. Welling, "Sigma delta quantized networks," *arXiv preprint arXiv:1611.02024*, 2016.

[20] P.-E. Novac, G. B. Hacene, A. Pegatoquet, B. Miramond, and V. Gripon, "Quantization and deployment of deep neural networks on microcontrollers," *Sensors*, vol. 21, no. 9, p. 2984, 2021.

[21] S. K. Esser, J. L. McKinstry, D. Bablani, R. Appuswamy, and D. S. Modha, "Learned step size quantization," *arXiv preprint arXiv:1902.08153*, 2019.

[22] R. Krishnamoorthi, "Quantizing deep convolutional networks for efficient inference: A whitepaper," *arXiv preprint arXiv:1806.08342*, 2018.

[23] Y. Yu, R. Hira, J. N. Stirman, W. Yu, I. T. Smith, and S. L. Smith, "Mice use robust and common strategies to discriminate natural scenes," *Scientific reports*, vol. 8, no. 1, pp. 1–13, 2018.

[24] K. Simonyan and A. Zisserman, "Two-stream convolutional networks for action recognition in videos," *arXiv preprint arXiv:1406.2199*, 2014.

[25] K. Soomro, A. R. Zamir, and M. Shah, "Ucf101: A dataset of 101 human actions classes from videos in the wild," *arXiv preprint arXiv:1212.0402*, 2012.

[26] J. Choi, Z. Wang, S. Venkataramani, P. I.-J. Chuang, V. Srinivasan, and K. Gopalakrishnan, "Pact: Parameterized clipping activation for quantized neural networks," *arXiv preprint arXiv:1805.06085*, 2018.

[27] S. Zhou, Y. Wu, Z. Ni, X. Zhou, H. Wen, and Y. Zou, "Dorefa-net: Training low bitwidth convolutional neural networks with low bitwidth gradients," *arXiv preprint arXiv:1606.06160*, 2016.

[28] M. A. Khoei, A. Yousefzadeh, A. Pourtaherian, O. Moreira, and J. Tapson, "Sparnet: Sparse asynchronous neural network execution for energy efficient inference," in *2020 2nd IEEE International Conference on Artificial Intelligence Circuits and Systems (AICAS)*, pp. 256–260, IEEE, 2020.

[29] S. Huang, A. Ankit, P. Silveira, R. Antunes, S. R. Chalamalasetti, I. El Hajj, D. E. Kim, G. Aguiar, P. Bruel, S. Serebryakov, *et al.*, "Mixed precision quantization for reram-based dnn inference accelerators," in *2021 26th Asia and South Pacific Design Automation Conference (ASP-DAC)*, pp. 372–377, IEEE, 2021.

[30] M. Kang, H. Kim, H. Shin, J. Sim, K. Kim, and L.-S. Kim, "S-flash: A nand flash-based deep neural network accelerator exploiting bit-level sparsity," *IEEE Transactions on Computers*, 2021.

[31] P. Vijayan, "Temporal Delta Layer." http://resolver.tudelft.nl/uuid:0806241d-9037-4094-a197-6e65d6482f2b.

[32] O. Vermesan and M. Diaz Nava (Eds), Intelligent Edge-Embedded Technologies for Digitising Industry ISBN: 9788770226103, River Publishers, Gistrup, Denmark, 2022.

4

An End-to-End AI-based Automated Process for Semiconductor Device Parameter Extraction

Dinu Purice[1], Matthias Ludwig[2], and Claus Lenz[1]

[1]Cognition Factory GmbH, Germany
[2]Infineon Technologies AG, Germany

Abstract

In this work, we present an automated AI-supported end-to-end technology validation pipeline aiming to increase trust in semiconductor devices by enabling a check of their authenticity. The high revenue associated with the semiconductor industry makes it vulnerable to counterfeiting activities potentially endangering safety, reliability and trust of critical systems such as highly automated cars, cloud, Internet of Things, connectivity, space, defence and supercomputers [7]. The proposed approach combines semiconductor device-intrinsic features extracted by artificial neural networks with domain expert knowledge in a pipeline of two stages: (i) a semantic segmentation stage based on a modular cascaded U-Net architecture to extract spatial and geometric information, and (ii) a parameter extraction stage to identify the technology fingerprint using a clustering approach. An in-depth evaluation and comparison of several artificial neural network architectures has been performed to find the most suitable solution for this task. The final results validate the taken approach, with deviations close to acceptable levels as defined by existing standards within the industry.

Keywords: Semantic segmentation, image processing, hardware trust, physical inspection of electronics, AI, ML, deep learning, supervised learning, convolutional neural networks, computer vision.

DOI: 10.1201/9781003377382-4

4.1 Introduction

Automation is one of the key parameters industries can approach to strengthen quality and lower overall costs. The improved availability of data and the mainstream application of approaches relying on artificial intelligence (AI) pushes industries towards the adaption of these AI methods. Nonetheless, practical implementations of these often seem to fail due to inflated expectations. Via a use-case from the semiconductor industry, we show various practical ways to overcome these potential pitfalls.

The recently introduced European Chips act recognises the paramount importance of the semiconductor industry within the global economy. The market for integrated electronics was at \$452.25B in 2021 and is expected to grow to \$803.15B in 2028 [8]. The high revenue potential causes extreme cost pressure and a highly competitive market. Consequently, since decades, the semiconductor industry is driven to automation along the complete value chain. One way to differentiate from competitors is through the utilisation of AI-powered manufacturing enhancements which have the potential to gain \$35B - \$40B annually over the entire industry [10]. Yet, not only manufacturing yields the potential to benefit from the industries push towards AI. The methods also offer the chance to be used for trust generation. In the aforementioned staggering market, rogues also aim to catch their share through counterfeiting, i.e. cloning, remarking, overproducing, or simply reselling of used parts [9]. This leads to the use case discussed throughout this work: via physical inspection and a fully integrated AI flow we present a fully automated assessment of the technological properties of a device. The idea for such a pipeline has already been introduced in [15] where it is argued that through a subsequent analysis of the cross-sections, the authenticity of the manufacturing technology can be validated. Relevant features in this case include geometric shapes and dimensions of the constituent structures, as well as material-related properties. Each technology can be interpreted as an individual fingerprint, such that deviations from specifications can be reported as suspicious. This work will focus on the end-to-end application aspects of the use case and includes following contributions:

- We will introduce an end-to-end, fully automated flow for semiconductor device technological parameter extraction by image segmentation and pattern recognition as an exemplary industrial use-case.
- We introduce our methodology that is tailored to the requirements of the use case. This includes an image segmentation approach which is

constituted of a set of specialised U-net cascades, class-specific loss functions, and an evolution-based training approach.

- The advantages of our design-decisions are quantitatively compared to similar state-of-the-art approaches and important lessons learned – transferable to other use-cases – are summarised.

Related work: The demand for measuring structures and critical dimensions within semiconductor devices is ever-increasing. While manufacturing relies mostly on in-line metrology, a further possibility is the post-production measurement. The databases are oftentimes big and automating of these flows is vital. A first template-based approach has been shown in [30]. This work relies on template matching and pattern recognition for the extraction of profile parameters. Furthermore, in a previous work [15], we have proposed how the flow can be utilised for the detection of counterfeit electronics [9] by comparing the extracted parameters against a database of known parameters.

The prospect of (semi)-automation of industrial processes through the use of machine learning-based (ML) methods is further gaining traction due to recent advancements in the field of ML and the uncovering of its unprecedented feature extraction and generalisation capabilities. Further accelerated due to the abundance of data, the "smartisation" of industrial processes through ML techniques has been conceived as the fourth industrial revolution [6].

The data set involved in this application bears two important characteristics: it consists of grey-valued images, and more importantly has a very limited availability of annotated data. The same characteristics are typically observed in medical applications, in dealing with images produced by computed tomography (CT), cone beam computed tomography (CBCT), as well as magnetic resonance imaging (MRI), ultrasound, X-ray, all data types being scarcely available to the public due to the confidential nature of medical data. Nevertheless segmentation tasks have been successfully tackled by ML-based methods, and in particular deep learning approaches which were proven to satisfy the high accuracy requirements typical to applications in the medical field. Of particular note in this context is the work of Ronneberger *et al.* [22] with the introduction of the U-net, a symmetric network consisting of a encoding and a decoding arm which was proved to possess high generalisation capabilities even on relatively small data sets. The progress was further accentuated after the debut of Dice-based loss functions, first introduced by Milletari *et al.* [17], which have been proven to outperform existing alternatives in the analysis of highly skewed data. Based

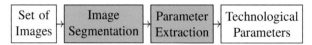

Figure 4.1 Overview of the architecture.

on the above-mentioned innovations, both supervised and unsupervised deep learning-based approaches have been constantly expanding within different use cases in the medical field, as shown by the works of Kawula *et al.* [11], Wang *et al.* [3] or Altaf *et al.* [2].

The following chapters describe the two paramount steps of this application, namely the Image Segmentation and the Parameter Extraction stages, respectively. Both stages are currently being fine-tuned and validated to ensure compliance with industry-defined standards of operation.

4.2 Semantic Segmentation

4.2.1 Proof of Concept and Architecture Overview

As a first step of development a benchmark stage was conducted, with the goal of determining the viability of an AI-based approach to scanning electron microscope (SEM) image segmentation and identify the most suitable architecture for the task. Considering that both the industrial sector and the academic sector lack openly available annotated semiconductor cross-section SEM data, a custom data set was assembled and labelled. The data set consists of 1024 by 685 grey-valued images, obtained at Infineon Technologies AG's failure analysis laboratories and represent technology nodes from 500 nm to approximately 40 nm with copper and Al-Tu technologies included. Devices with less than one metal layer (e.g. discrete transistors) were excluded. The image sources are state-of-the-art SEMs available in semiconductor failure analysis laboratories. For the purpose of this stage 202 images were manually sampled and labelled.

The images were annotated with 5 relevant labels of interest, namely "metal", "VIA", "lateral isolation", "poly", and "deep trench isolation" [25], each bearing features important in the process of technology identification. The selected features imbue the following purposes within a semiconductor device:

- **Metal:** Low resistance metallic connections between devices. Several metallisation layer can be stacked over each other to route inter-device connections.

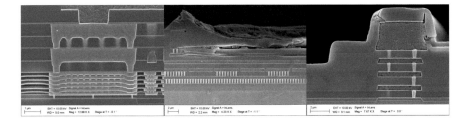

Figure 4.2 Examples showcasing different semiconductor technologies

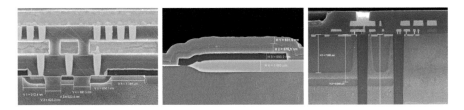

Figure 4.3 Examples of labelled data showcasing the different ROIs: green – VIA; yellow – metal; teal – lateral isolation; red – poly; blue – deep trench isolation

- **Vertical interconnect access (VIA) / contact:** Low ohmic interconnections between different metallisation layers (VIA) or between devices and the lowest metallisation layer.
- **Lateral isolation (shallow trench isolation):** Electrical lateral isolation between devices with a dioxide trough a *shallow* deposition.
- **Deep trench isolation:** Trenches for lateral isolation with a high depth-width ratio. Mostly found in analogue integrated circuits.
- **Poly:** Poly-crystalline silicon which is used as gate electrode.

For the benchmark stage however only two regions of interest (ROIs) were selected, namely "VIA" and "metal". The two ROIs strongly differ in terms of size and quantity, with the pixel-wise class-distribution of the "metal" objects representing 13.61%, while "VIA" objects being more numerous but at the same time smaller, taking up 2.5%. Therefore, they reflect the two important properties of the expected data: high variability and high skewness.

As it can be seen in Fig. 4.4 there is a strong overlap in intensity between the various regions of interest, yielding classical segmentation methods such as thresholding [21], region-growing [20], watershed [18] and k-means clustering [19] ineffective. Instead, an effective segmentation process requires

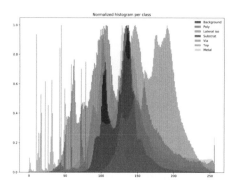

Figure 4.4   Histograms of the investigated data grouped by label of interest

domain-expert knowledge – thus encouraging the use of deep learning-based methods capable of extracting spatial and semantic features. Several network architectures were selected as candidates, based on their respective performance in similar segmentation tasks. An overview of each candidate network architecture is presented below:

- **U-net** [22]

Introduced by Ronneberger *et al.* [22] as a solution for biomedical image segmentation, this architecture has been shown to perform reasonably well even when trained with small amounts of data. It consists of an down-sampling encoder and an up-sampling decoder arm enabling efficient spatial context capture. The arms are connected with skip connection which accelerate convergence during training and combat vanishing gradients. The U-net achieved an averaged Dice score of 0.76 on the test subset.

- **Feature Pyramid Network (FPN)** [13]

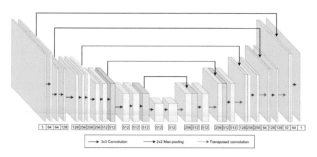

Figure 4.5   Overview of the U-net architecture [24]

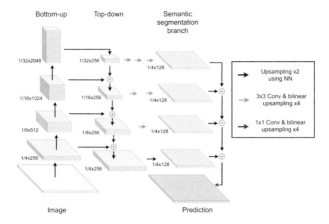

Figure 4.6 Overview of the FPN architecture [24]

The FPN follows a top-down approach with skip connections, similar to the previously mentioned U-net. However instead of using the final output as the prediction, the FPN makes predictions for each stage (see Fig. 4.6) thus combining semantically strong low-resolution features with semantically weaker high-level features. An additional segmentation branch is used to then merge the information from all levels into a single output. The FPN obtained an averaged Dice score of 0.71 on the test subset

- **Gated-Shape Convolutional Neural Network (GSCNN)** [27]

The GSCNN employs a two-stream architecture, with the shape-related features focused in a dedicated stream that works in parallel to the standard

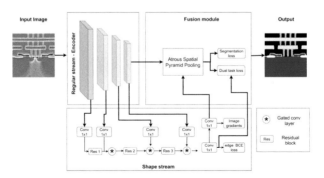

Figure 4.7 Overview of the GSCNN architecture [24]

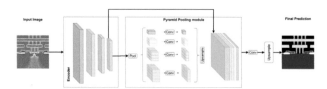

Figure 4.8 Overview of the PSPNet architecture [24]

encoder. A key characteristic of this architecture is the use of gated convolutional layers, which connect intermediate layers of both streams, facilitating the transfer of information from the encoder to the shape stream while filtering irrelevant information. The information of both streams is then combined within the fusion stage using an Atrous Spatial Pyramid Pooling module (ASPP). An averaged Dice score of 0.74 on the test subset was obtained by the GSCNN.

- **Pyramid Scene Parsing Network (PSPNet)** [31]

The PSPNet architecture makes use of a Pyramid Pooling Module (PPM) to capture rich context information from the output of the encoder arm. The capture is done through fusion of the network's four pyramid scales, as seen in Fig. 4.8. An averaged Dice score of 0.69 on the test subset was obtained using the PSPNet architecture.

- **Siamese network** [16]

The Siamese network presents another approach to combine the features extracted at low-resolution and high-resolution levels, namely through a two step approach. The first step operates on the whole, down-sampled image, and outputs a coarse segmentation map. As a second step the segmentation

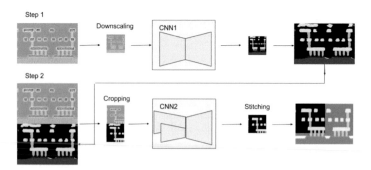

Figure 4.9 Overview of the Siamese network architecture [12]

map is then fed into a Siamese network containing two encoders (as show in Fig. 4.9), with the original high resolution image going through the other encoder in patches. Finally the decoder stitches together the patches, obtaining a segmentation map at the same resolution as the input image. The Siamese network reached an averaged Dice score of 0.78 on the test subset.

4.2.2 Implementation Details and Result Overview

To complete the benchmark stage, each network architecture was trained 5 times on random pre-sampled splits of the data set (60% training, 20% validation, 20% test). The resulting Dice scores (averaged over the 5 tries and the 2 labels of interest) and their respective spread is presented in Fig. 4.10 below. All experiments were ran on a server equipped with Intel Core i9-9940x (14 Cores, 3,30GHz), 4 RTX 5000 GPUs and 128 GB RAM.

The two best performing approaches are the Siamese (with a mean Dice score of 0.78) and the U-net (with a mean Dice score of 0.76). The performance of the Siamese approach can be explained by the two steps

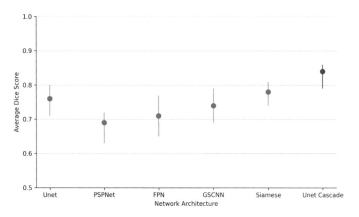

Figure 4.10 Average Dice Scores (blue) and spread (green) per investigated network architecture, along with the final chosen architecture (red)

Table 4.1 Obtained Dice Scores for each showcased network architecture

Architecture	U-net	PSPNet	FPN	GSCNN	Siamese
Average DSC	0.76	0.69	0.71	0.74	0.78
DSC range	0.71 - 0.80	0.63 - 0.72	0.65 - 0.77	0.69 - 0.79	0.74 - 0.81

analysis employed by this method, which segments firstly in low resolution therefore with a larger perceptive field, followed by a second step analysing in a higher resolution, with the downside of having a lower perceptive field. On the other hand, the U-net architecture obtained similar results with much lower resource consumption during training and inference.

Based on this performance, a branched U-net cascade was chosen as the preferred architecture, combining both the two step analysis at different resolution levels, as well as the generalisation power associated with the U-net. The chosen architecture consists of independent branches targeting each ROI. For each branch, a 2D U-net takes the down-sampled image as input and produces an intermediate, rough segmentation, which is then up-sampled to the dimensions of the original input image. The intermediate segmentation is then aggregated with the original high-resolution input image to be fed into a 3D U-net (as introduced by Milletari *et al.* [17]), which then outputs a high-resolution segmentation map. Some practical advantages of such a modular architecture are the possibility to update each branch individually in case of additional data being available, as well as to allow scaling up with additional branches targeting new labels without having to update each branch. An overview of the described architecture for a given branch is presented in Fig. 4.11. Repeating the experiment in benchmark conditions has yielded an averaged Dice score of 0.84 (shown in red in Fig. 4.10), outperforming all the other candidate architectures.

Typical to deep learning applications with limited data sets and despite the use of data augmentation techniques, overfitting was proven to be an issue. This could be clearly seen in the discrepancy between the Dice scores on the train and test subsets respectively. Having chosen an architecture for the fine-tuning stage, additional effort was invested in expanding the data set from 202 to 2192 images. For this stage of the application all five previously mentioned labels of interest were trained on.

Due to the relatively large number of hyper-parameters to be tuned a population-based training method was used, consisting of two evolution phases: exploration and exploitation. During the exploration phase the networks are trained with randomly sampled hyper-parameters sets. During the following exploitation phase the best performing sets of hyper-parameters are identified, and new sets are sampled in close proximity within the hyper-parameter space.

Although Dice loss has proven itself effective in segmentation tasks, the high skewness and variability as well as low availability of data require additional compensatory mechanisms. For this purpose several alternative

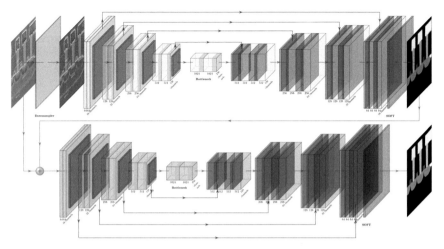

Figure 4.11 An overview of the U-net cascade architecture, consisting of a 2D U-net (top) and a 3D U-net (bottom) which takes as input the high resolution input image stacked with the output segmentation of the first stage

loss functions were investigated as hyper-parameters, including Focal Tversky loss [1], Combo Loss [26], Unified Focal Loss (LogCoshDSC) [29]. Training experiments indicated that the loss function is the paramount hyper-parameter, having the most impact upon the resulting accuracy of the network. Furthermore, different labels have been shown to benefit differently from each loss function. For example the network trained on the "metal" label, which has the highest pixel-wise distribution of all classes and typically large structures on each image, performed best when trained using the LogCoshDSC loss. At the same time the "VIA" and "poly" labels, both with a very low pixel-wise distribution ($< 2.5\%$) were segmented best by networks trained with the Focal Tversky loss. The Combo loss on the other hand was most effective for the networks targeting the "lateral iso" and "deep trench" labels, which have an average pixel-wise distribution but are the most difficult to identify visually.

The average Dice scores obtained on the test set for each label of interest are presented in the table below.

The "metal" and "VIA" labels obtained the highest Dice scores, with a substantial increase in accuracy of about 10% compared to the benchmark stage. Also of particular note is the "deep trench" case. Despite being the class with the lowest pixel-wise distribution, only appearing in 58 of the images,

Table 4.2   Averaged Dice Scores for each label of interest

Label	Metal	VIA	Poly	Lateral iso	Deep trench
Loss function	LogCosh	Foc. Tversky	Foc. Tversky	Combo	Combo
Average DSC	0.93	0.91	0.88	0.82	0.76

the proposed network architecture was able to segment it with reasonable accuracy to make use of the extracted information.

4.3 Parameter Extraction

The process following the semantic image segmentation is the extraction of the technological device parameters. The overview of the algorithmic approach is shown in algorithm 1. The inputs are the image meta-data – with the sole relevant information being the pixel size per image – and the segmented image. In a first step the segmented are written to polygon while retaining their class-labels. Subsequently, the polygons of every class (C) are retrieved. From this set of polygons, polygons below a statistically evaluated threshold (area of a polygon instance lower than five times the mean of a polygon instances within this class) are removed from the list From these *cleaned* polygons, the centroids of the single objects are computed which are utilised for clustering. The customised clustering method is shown in table 4.3.

Algorithm 1 Parameter extraction approach

    ▷ Input: Segmented image and corresponding meta-data

1: **function** PARAMETEREXTRACTION($Image, Meta$)
2:    $Polygons_C \leftarrow GetPolygons(Image, Meta)$
3:    **for all** $Classes$ **do**
4:        $Polygons_C \leftarrow GetPolygonsOfSingleClass(Image)$
5:        $Polygons_C \leftarrow CleanPolygons(Polygons_C)$
6:        $Polygons_{C,i} \leftarrow ClusterPolygonsVertically(Polygons_C.Centroid)$
7:        **for all** $Clusters$ **do**
8:            $Polygons_{C,i,j} \leftarrow ClusterPolygonsHorizontally(Polygons_{C,i})$
9:            $Parameters_{C,i} \leftarrow ExtractParameters(Polygons_{C,i,j})$
10:        **end for**
11:    **end for**
12:    **return** $Parameters$
13: **end function**

Table 4.3 Utilised cluster evaluation techniques [14]. Notation: n: number of objects in dataset; c: centre of data-set; NC: number of clusters; C_i: the i-th cluster; n_i: number of objects in C_i; c_i: centre of C_i; W_k: the within-cluster sum of squared distances from cluster mean; $W*_k$ appropriate null reference; B reference data-sets

Method	Definition	Value
CH [4]	$$\dfrac{\sum_i n_i \cdot d^2(c_i, c)/(NC - 1)}{\sum_i \sum_{x \in C_i} d^2(x, c_i)/(n - NC)}$$	Elbow
Gap [28]	$$log\left(\dfrac{(\prod W_{kb}^*)^{1/B}}{W_k}\right)$$	Elbow
DB [5]	$$\dfrac{1}{NC} \cdot \sum_i \left\{\dfrac{\dfrac{1}{n_i}\sum_{x \in C_i} d(x, c_i) + \dfrac{1}{n_j}\sum_{x \in C_j} d(x, c_j)}{d(c_i, c_j)}\right\}$$	Min
DB^2 [5]	$$\dfrac{1}{NC} \cdot \sum_i \left\{\dfrac{\dfrac{1}{n_i}\sum_{x \in C_i} d(x, c_i) + \dfrac{1}{n_j}\sum_{x \in C_j} d(x, c_j)}{d(c_i, c_j)} \cdot i^2\right\}$$	Min
Sil. [23]	$$\dfrac{1}{NC} \sum_i \left\{\dfrac{1}{n_i}\sum_{x \in C_i} \dfrac{b(x) - a(x)}{max[b(x), a(x)]}\right\}$$	Max

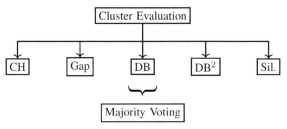

Figure 4.12 Utilised cluster evaluation techniques.

Different cluster evaluation techniques – namely Calinski-Harabasz (CH) [4], gap [28], Davies-Bouldin (DB) [5], a custom squared Davies-Bouldin (DB^2) [5], and silhouette (Sil.) [23] – are conducted on one-dimensional feature vectors which are constituted of the y-share of the centroid coordinates. The previous clustering is done via trough k-means clustering while the k is kept – adapted to the use case – between *2* and *10*. For the different evaluation techniques optimal number of clusters (k) are reported through different metrics (minimum, maximum, elbow). This computationally costly approach is suitable for the use case since the vectors are one-dimensional and the total number of polygon objects to be evaluated is relatively small (< 100). In the final step of the vertical clustering, the optimal number of clusters is inferred through a majority voting among the individual evaluation techniques.

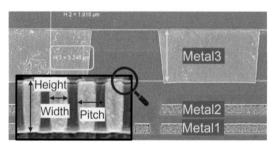

Figure 4.13   Example cross-section image with annotated metal and contact/VIA features

Since the polygon objects are now vertically assigned, a clustering in the horizontal dimension is the next step. The procedure is the same as previously discussed for the vertical clustering. For the vertically and horizontally clustered elements, the technological, geometrical parameters can be inferred. These are illustrated via figure 4.13 for the metal and VIA classes. The vertical height is determined for metallisation layers and height, width, and pitches for the interconnecting contact and VIA layers. After the polygons objects are assigned to classes, these attributes can be calculated through trivial mathematical operation. Height is the difference of the bounding box maximum and minimum in vertical dimension. Width is the difference of the bounding box in horizontal dimension, and pitches are the differences of the x-coordinate of the centroid of two adjacent polygon objects. The values are respectively averaged within all classes.

An example is shown for the VIA class through the example in figure 4.14. After segmentation of the grey-scale image, the individual segmented classes are inferred into polygon objects. Here the VIA class is exemplified. The vertical clustering process is shown through the two right images. The dendrogram visualises the linkage of the different clusters which are subsequently optimised via discussed approach. The optimum number of clusters are shown in the bottom right figure. An evaluation techniques report an optimum of four (different values constitute the optimum) clusters. Following this, these four are subsequently clustered in horizontal dimension and respectively geometrically inferred. Following results were obtained for this example (besides the absolute values, the relative deviation to a manual measurement is given):

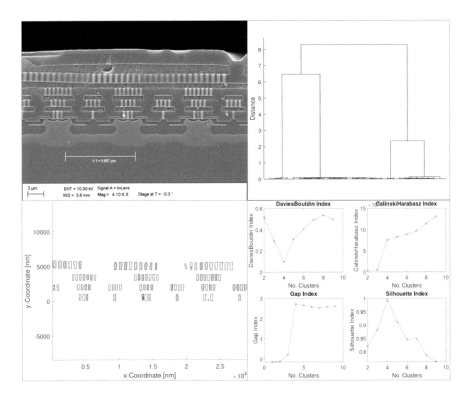

Figure 4.14 Example cross-section image (upper left). The polygonised VIA objects are shown (lower left). A dendrogram is shown for the relative distances of the y-coordinates of the single objects (upper right). Finally, the results of the utilised cluster evaluation techniques are presented (lower right).

- Contact (**h**eight, **w**idth, **p**itch): 942 nm (+9.66%), 319 nm (+5.98%), 525 nm (+12.42%)
- VIA1 (h, w, p): 870 nm (+26.82%), 319 nm (n.a.), 545 nm (n.a.)
- VIA2 (h, w, p): 898 nm (+14.83%), 319 nm (n.a.), 542 nm (n.a.)
- VIA3 (h, w, p): 1086 nm (+16.27%), 434 nm (n.a.), 750 nm (n.a.)

Within the technology validation use case, the inferred technological features are tested against the designed and manufactured technological properties. This is computed via multi-dimensional distance matching (e.g. Euclidean, rectilinear distance) of both vectors. The validation accuracy depends on several different factors which are the segmentation quality, parameter extraction accuracy and image acquisition completeness. Experiments have shown that the current automated end-to-end flow reaches 75% accuracy for previously known Al-Tu technologies. Improvement is necessary for copper (Cu) technologies which are more complex to segment. According to existing procedures within the industry, deviations of less than 5% for pitches and deviations of less than 25% for all other geometrical measurements compared to a ground truth, i.e. the designed technology parameters are acceptable. The same requirements have been used as a benchmark for the validation of this application. The high deviations are a consequence attributed to process variances during device manufacturing and de-processing. Presented image shows a single frame which was acquired in a sub-optimal zoom level for measuring discussed features. Yet, almost all requirements were achieved. In summary it can be stated that the proof-of-concept presented in this work displays strong potential to satisfy existing industrial requirements, especially when adequate zooms levels are chosen for the particular technological parameters.

4.4 Conclusion

The settings for AI implementation in an industrial setting are often completely different from consumer applications. Data being scarce the design of productive AI application is forcibly *data-driven*, or more specifically *data-adapted*. Industrial parameters are manifold, and the requirements typically impose the need to automate, improve, or even enable new processes. To make an AI-based solution *viable* these requirements must be met. In this work, we have shown through an end-to-end technology demonstrator – incorporating deep learning and cluster evaluation – showcasing the automation of semiconductor technology identification based on SEM cross-section analysis. A comparison of different convolutional neural network architectures was presented, and a candidate best suited for the SEM segmentation task was drafted. The proposed candidate architecture represents a cascade of 2D and 3D Unets, arranged in branches each dedicated to a single label of interest. Following a pragmatic perspective, a modular design is proposed, ensuring scalability and ease-of-maintenance. Trained on a custom-created data set of 2192 images, the proposed architecture obtained Dice scores in the range of 0.76-0.93 for labels of different complexity, arguing in favour of the employment of supervised deep learning-based methods even in applications with strongly limited amounts of available labelled data. Based on the obtained results, a parameter extraction algorithm is proposed, aimed at exploiting the obtained segmentation maps with the purpose of identifying and validating the technology of the investigated semiconductor devices. The obtained results were in the range of ground truth measurements with deviations in an acceptable measuring range. The potential for narrowing down these uncertainty ranges were outlined.

4.5 Future Work

Following the development and validation steps described above, a production test stage will determine the potential of the segmentation component of the process to be used in other applications of semiconductor analysis. Aside from a high degree of automation and the mandatory fulfilment of functional requirements, industry has established high thresholds for non-functional requirements. Maintainability, system up time, extensibility, usability, or updateability are just some of the potential requirements across different industries. Such requirements are addressed by specialised frameworks such as Ray and TorchServe. Combined with the advantages of a modular architecture, they enable the possibility to update each network with virtually no down-time. Additional investigations are conducted in the expansion of the data augmentation pipeline, with the goal of increasing the exploitation of the available data set regardless of its relatively small size.

Acknowledgment

This work is conducted under the framework of the ECSEL AI4DI "Artificial Intelligence for Digitising Industry" project. The project has received funding from the ECSEL Joint Undertaking (JU) under grant agreement No 826060. The JU receives support from the European Union's Horizon 2020 research and innovation programme and Germany, Austria, Czech Republic, Italy, Latvia, Belgium, Lithuania, France, Greece, Finland, Norway.

References

[1] N. Abraham and N. Mefraz Khan. A novel focal tversky loss function with improved attention u-net for lesion segmentation. *2019 IEEE 16th International Symposium on Biomedical Imaging (ISBI 2019)*, pages 683–687, 2019.

[2] F. Altaf, S. M. S. Islam, N. Akhtar, and N. Khalid Janjua. Going deep in medical image analysis: Concepts, methods, challenges, and future directions. *IEEE Access*, 7:99540–99572, 2019.

[3] S. Budd, E. C. Robinson, and B. Kainz. A survey on active learning and human-in-the-loop deep learning for medical image analysis. *Medical Image Analysis*, 71:102062, 2021.

[4] T. Caliński and J. Harabasz. A dendrite method for cluster analysis. *Communications in Statistics - Theory and Methods*, 3(1):1–27, 01 1974.

[5] D. L. Davies and D. W. Bouldin. A cluster separation measure. *IEEE Transactions on Pattern Analysis and Machine Intelligence*, PAMI-1(2):224–227, 1979.

[6] A. Diez-Olivan, J. Del Ser, D. Galar, and B. Sierra. Data fusion and machine learning for industrial prognosis: Trends and perspectives towards industry 4.0. *Information Fusion*, 50:92–111, 2019.

[7] European Commision. A chips act for europe.

[8] Fortune Business Insights. Semiconductor market size, share & covid-19 impact analysis, 2021-2028. FBI102365.

[9] U. Guin, K. Huang, D. DiMase, J. M. Carulli, M. Tehranipoor, and Y. Makris. Counterfeit integrated circuits: A rising threat in the global semiconductor supply chain. *Proceedings of the IEEE*, 102(8):1207–1228, 2014.

[10] S. Göke, K. Staight, and R. Vrijen. Scaling ai in the sector that enables it: Lessons for semiconductor-device makers.

[11] M. Kawula, D. Purice, M. Li, G. Vivar, S.-A. Ahmadi, K. Parodi, C. Belka, G. Landry, and C. Kurz. Dosimetric impact of deep learning-based ct auto-segmentation on radiation therapy treatment planning for prostate cancer. *Radiation Oncology*, 17, 01 2022.

[12] G.-M. Konnerth. Exploring application-oriented methods to improve cnn-based segmentation of sem microchip images. Master's thesis, Technical University of Munich, 2020.

[13] X. Li, T. Lai, S. Wang, Q. Chen, C. Yang, R. Chen, J. Lin, and F. Zheng. Weighted feature pyramid networks for object detection. In *2019 IEEE Intl Conf on Parallel Distributed Processing with Applications, Big Data Cloud Computing, Sustainable Computing Communications, Social Computing Networking (ISPA/BDCloud/SocialCom/SustainCom)*, pages 1500–1504, 2019.

[14] Y. Liu, Z. Li, H. Xiong, X. Gao, and J. Wu. Understanding of internal clustering validation measures. In *2010 IEEE International Conference on Data Mining*. IEEE.

[15] M. Ludwig, B. Lippmann, A.-C. Bette, and C. Lenz. Demo: A fully automated process for semiconductor technology

analysis through SEM cross-sections. In *25th International Conference on Pattern Recognition (ICPR)*.

[16] K. Martin, N. Windunga, S. Sani, S. Massie, and J. Clos. A convolutional siamese network for developing similarity knowledge in the selfback dataset, 2017.

[17] F. Milletari, N. Navab, and S.-A. Ahmadi. V-Net: Fully Convolutional Neural Networks for Volumetric Medical Image Segmentation. In *2016 Fourth International Conference on 3D Vision (3DV)*, pages 565–571. IEEE, 10 2016.

[18] L. Najman and M. Schmitt. Watershed of a continuous function. *Signal Processing*, 38(1):99–112, 1994. Mathematical Morphology and its Applications to Signal Processing.

[19] D. Nameirakpam, K. Singh, and Y. Chanu. Image segmentation using k -means clustering algorithm and subtractive clustering algorithm. *Procedia Computer Science*, 54:764–771, 12 2015.

[20] R. Nock and F. Nielsen. Statistical region merging. *IEEE Transactions on Pattern Analysis and Machine Intelligence*, 26(11):1452–1458, 2004.

[21] N. Otsu. A threshold selection method from gray level histograms. *IEEE Transactions on Systems, Man, and Cybernetics*, 9:62–66, 1979.

[22] O. Ronneberger, P. Fischer, and T. Brox. U-net: Convolutional networks for biomedical image segmentation. In Nassir Navab, Joachim Hornegger, William M. Wells, and Alejandro F. Frangi, editors, *Medical Image Computing and Computer-Assisted Intervention – MICCAI 2015*, pages 234–241, Cham, 2015. Springer International Publishing.

[23] P. Rousseeuw. Silhouettes: A graphical aid to the interpretation and validation of cluster analysis. *J. Comput. Appl. Math.*, 20(1):53–65, November 1987.

[24] F. Schiegg. Boundary detection and semantic segmentation of sem-images. Master's thesis, Technical University of Munich, 2020.

[25] S. M. Sze and K. K. Ng. *Physics of Semiconductor Devices*. Wiley, 2006.

[26] S. A. Taghanaki, Y. Zheng, S. K. Zhou, B. Georgescu, P. Sharma, D. Xu, D. Comaniciu, and G. Hamarneh. Combo loss: Handling input and output imbalance in multi-organ segmentation. *Computerized Medical Imaging and Graphics*, 75:24–33, 2019.

[27] T. Takikawa, D. Acuna, V. Jampani, and S. Fidler. Gated-scnn: Gated shape cnns for semantic segmentation. In *2019 IEEE/CVF International Conference on Computer Vision (ICCV)*, pages 5228–5237, 2019.

[28] R. Tibshirani, G. Walther, and T. Hastie. Estimating the number of clusters in a data set via the gap statistic. 63(2):411–423.

[29] M. Yeung, E. Sala, C.-B. Schönlieb, and L. Rundo. A mixed focal loss function for handling class imbalanced medical image segmentation. *ArXiv*, abs/2102.04525, 2021.

[30] X. Zhang, Z. Fu, Y. Huang, A. Lin, Y. Shi, and Y. Xu. Effective method to automatically measure the profile parameters of integrated circuit from SEM/TEM/STEM images. In *2017 China Semiconductor Technology International Conference (CSTIC)*. IEEE, 3 2017.

[31] H. Zhao, J. Shi, X. Qi, X. Wang, and J. Jia. Pyramid scene parsing network. In *2017 IEEE Conference on Computer Vision and Pattern Recognition (CVPR)*, pages 6230–6239, 2017.

5

AI Machine Vision System for Wafer Defect Detection

Dmitry Morits, Marcelo Rizzo Piton, and Timo Laakko

VTT Technical Research Centre of Finland Ltd, Finland

Abstract

Surface defects generated during semiconductor wafers processing are among the main challenges in micro- and nanofabrication. The wafers are typically scanned using optical microscopy and then the images are inspected by human experts. That tends to be a quite slow and tiring process. The development of a reliable machine vision-based system for correct identification and classification of wafer defect types for replacement of manual inspection is a challenging task, due to the variety of possible defects. In this work we developed a machine vision system for the inspection of semiconductor wafers and detection of surface defects. The system integrates an optical scanning microscopy system and an AI algorithm based on the Mask R-CNN architecture. The system was trained using a dataset of microscopic images of wafers with Micro Electro-Mechanical Systems (MEMS), silicon photonics and superconductor devices at different fabrication stages including surface defects. The achieved accuracy and detection speed makes the system promising for cleanroom applications.

Keywords: AI, machine vision, semiconductor wafer, defect detection, convolutional neural network, Mask R-CNN.

5.1 Introduction and Background

One of the main challenges in micro- and nanofabrication is the identification and classification of surface defects. The defects are unavoidably generated

DOI: 10.1201/9781003377382-5

during processes such as chemical-mechanical polishing, photolithography, etching, diffusion and ion implantation, oxidation, metallization, and others [1][2]. The increasing complexity and density of semiconductor devices leads to an increase of the number of surface defects and dictates stricter requirements for defect detection. For example, contamination particles harmless for some design rules at the same time could be critical as the device dimensions grow smaller. The defect criteria are also varying in different locations of devices: for example, defects in a movable part or in the hermetic bond-sealing frame of a MEMS device are usually more severe than in secondary structures. Figure 5.1 illustrates microscopic images with surface defects generated during the microfabrication of different superconductor and semiconductor devices. Typical types of defects include particles, photoresist spots, edge defects, scratches, etc. It becomes evident that defect detection is an extremely important procedure, especially at the critical areas of the devices.

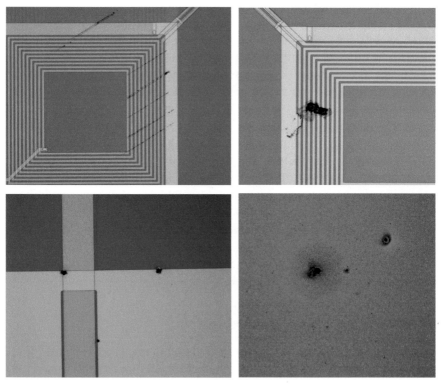

Figure 5.1   Examples of microscopic images of various superconductor and semiconductor devices with surface defects

VTT Micronova semiconductor fab is a Finnish national research infrastructure for micro-, nano- and quantum technology. The research areas include MEMS, photonic, quantum and other specialty components that can be used to create a wide range of sensors and devices. At VTT, the current visual inspection process of the wafer surface is manually performed by human experts. The wafers are scanned using optical microscopy, and then the images are inspected by the human experts. Since the inspection task requires extreme concentration, the time that an expert can perform the task is quite limited. Additionally, it tends to be a quite slow, tiring process and susceptible to human mistakes. Identification of defects by experts alone can potentially result in false identifications due to fatigue and lack of objectivity. The goal of this work is the development of a reliable machine vision-based system for the correct identification of wafer defects in the hope of replacing manual inspection. Moreover, this system would be directly integrated in the wafer inspection production line. Such a system would speed up the defect inspection, simplify the analysis and eventually help to improve the fabrication yield.

5.2 Machine Vision-based System Description

The general architecture of the developed machine vision system is shown in Figure 5.2. The wafers are inspected by a semi-automatic microscopy scanning system. In this work we tested both IJ 13 IR-inspector and Muetec CD3000 optical scanning system. The system produces a set of microscopic images, covering the full area of the wafer.

For the training of neural networks, we prepared an image dataset using microscopic images of wafers with MEMS, silicon photonics and superconductor devices at different fabrication stages including surface defects. The initial set included images of different resolutions and magnifications. First, we manually labelled the defects on each image and then cropped the areas with defects. The cropping allowed the increase of the dataset size and provided faster and more consistent training. Next, a data augmentation technique was used to increase the amount of data by adding slightly modified copies of already existing data, or newly created synthetic data from existing data. That procedure acts as a regularizer and helps to reduce overfitting when training a machine learning model [3]. In this case, the augmentation included mirror and rotation image transformation, as well as a change of the RGB spectre of the images. The full procedure of dataset preparation is

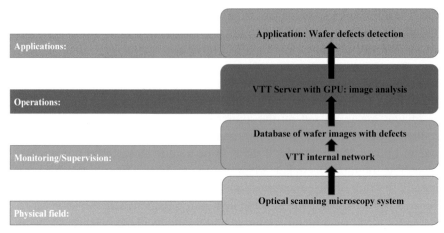

Figure 5.2 General architecture of the developed machine vision system

schematically shown in Figure 5.3. The dataset was split into training and validation sets, containing 935 and 165 images each.

Here we used a Convolutional Neural Network (CNN): a special type of deep learning algorithm, used primarily for image recognition and processing. CNNs are inspired by the organization of the animal visual cortex [4][5] and are designed to learn spatial hierarchies of features, from low- to high-level patterns. We developed an algorithm based on the Mask R-CNN architecture [6], which is a state-of-the-art algorithm for object detection - a computer vision technique that enables the identification and location of objects in an image or video. Mask R-CNN is the latest stage of evolution of CNNs, providing high detection accuracy. At the same time, it requires more computational resources compared to faster algorithms, such as YOLO [7]. Mask R-CNN consists of two stages. The first stage, called a Region

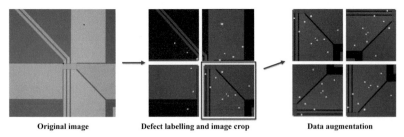

Original image   Defect labelling and image crop   Data augmentation

Figure 5.3 A scheme of the image dataset preparation, including labelling, cropping and data augmentation

Proposal Network, proposes candidate object bounding boxes. The second stage extracts features using Region of Interest Pool from each candidate box, then performs classification and bounding-box regression and outputs a binary mask for each Region. The ResNet-101 [8] convolutional backbone architecture was used for feature extraction over an entire image. The algorithm was optimized for so-called binary classification, which provides results in "defect vs background" format, without classification of defects, shown in Figure 5.4. The general comparison of the algorithm's performance to other object detection algorithms can be found in Refs [6] and [9].

Among the main requirements for the system are the functional suitability for defect detection, the integration of the scanning optical microscope and the server with the AI software, the usability for cleanroom users who are not familiar with the details of implementation, and the readability and visualization of the detection results for the users. The main KPIs for the system were: detection accuracy, time of processing a single image and evaluation by the cleanroom users from the points of usability and result readability. The AI algorithm based on the Mask R-CNN architecture passed several rounds of optimization and testing using microscopic images of various microelectronic devices.

There has been a significant progress in the application of deep learning techniques for wafer defect detection and classification [10]. The main innovation elements of this work compared to the state of the art is the integration of the algorithm with the scanning microscopy system, and training of the system using the dataset containing images of various devices at different stages of processing, instead of standard image databases available online. It allows the system to better distinguish between wafer defects and features of the devices and provides reliable detection of wafer defects for a wide range of semiconductor components.

To improve the system usability for the end-users, we implemented a Graphical User Interface adapted for cleanroom personnel not familiar with AI systems. The software was installed on a PC/server with NVIDIA Quadro RTX 5000 16GB GPU at the VTT Micronova cleanroom. Then the algorithm was integrated with the optical scanning microscopy system Muetec CD3000 by connection through the internal network. To improve the readability of the results, the system provides binary classification defect vs background with results available in both graphical and text formats. The feedback from the cleanroom experts helped in the improvement of system usability after several iterations of optimization. The testing results at the latest dataset with 192 images of 1600x1200px resolution and 5x optical magnification,

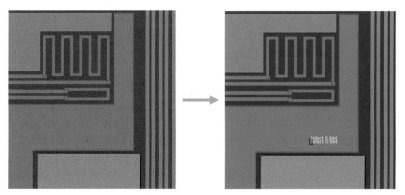

Figure 5.4   Example of binary classification of wafer defects: defect vs background

demonstrated 86% accuracy with a detection time of $1 \div 2$ seconds per image. The accuracy of the system is approximately on the same level as that of a human operator, although it also depends a lot on the experience of the operators and their tiredness. The experts estimated 86% accuracy as sufficient for applications at VTT cleanroom but mentioned that only about 15% of the detected defects were critical for wafer processing. Unfortunately, the criteria of a defect being critical or non-critical is very device-specific and cannot be easily generalized. After the system provides the detection results, the final decision on the importance of the defects for processing had to be made by the cleanroom experts.

Regarding the system scalability, in the current work we did not have the goal of moving towards smaller technology nodes, although such scaling might require utilization of faster neural networks, like one-stage YOLO detectors. In general, the main expected impact of the system development is the reduction of the overall working time required for wafer defect inspection. We believe that the system will help saving valuable working time of cleanroom experts, improve fabrication yield and reduce fabrication cost.

5.3 Conclusion

We developed a system for the detection of wafer surface defects. The system integrates an optical scanning microscopy system and an AI algorithm based on the Mask R-CNN architecture. The image dataset used for training and testing the system included microscopic images of wafers with MEMS, silicon photonics and superconductor devices at different fabrication stages including surface defects. The system demonstrated functional suitability for defect detection, high accuracy, and reasonable detection speed, making it suitable for potential cleanroom applications.

Acknownledgements

This work is conducted under the framework of the ECSEL AI4DI "Artificial Intelligence for Digitising Industry" project. The project has received funding from the ECSEL Joint Undertaking (JU) under grant agreement No 826060. The JU receives support from the European Union's Horizon 2020 research and innovation programme and Germany, Austria, Czech Republic, Italy, Latvia, Belgium, Lithuania, France, Greece, Finland, Norway.

References

[1] H. J. Queisser, E. E. Haller, "Defects in Semiconductors: Some Fatal, Some Vital", Science, 281, 945– 950, 1998.

[2] T. Yuan, W. Kuo, and S. J. Bae, "Detection of spatial defect patterns generated in semiconductor fabrication processes", *IEEE Trans. Semicond. Manuf.*, vol. 24, no. 3, pp. 392–403, Aug. 2011.

[3] A. Buslaev, V. I. Iglovikov,; E. Khvedchenya, A. Parinov, M. Druzhinin, A. A. Kalinin, "Albumentations: Fast and Flexible Image Augmentations", Information 11, 2, 2020. https://www.mdpi.com/2078-2489/11/2/125.

[4] S. Albawi, T. A. Mohammed, S. Al-Zawi, "Understanding of a convolutional neural network", *International Conference on Engineering and Technology (ICET), IEEE*, pp. 1-6, 2017.

[5] W. Liu, Z. Wang, X. Liu, N. Zeng, Y. Liu, F.E. Alsaadi, "A survey of deep neural network architectures and their applications", *Neurocomputing*, 234, pp. 11-26, 2017.

[6] K. He, G. Gkioxari, P. Dollár, R. Girshick, "Mask R-CNN". arXiv:1703.06870, 2018.

[7] M. Carranza-García, J. Torres-Mateo, P. Lara-Benítez, J. García-Gutiérrez, "On the Performance of One-Stage and Two-Stage Object Detectors in Autonomous Vehicles Using Camera Data", Remote Sens. 13, 89, 2021.

[8] K. He, X. Zhang, S. Ren, J. Sun, "Deep Residual Learning for Image Recognition", *Proceedings of the IEEE Conference on Computer Vision and Pattern Recognition*, 770-778, 2016.

[9] Z. Zhao, P. Zheng, S. Xu, X. Wu, "Object Detection With Deep Learning: A Review", *IEEE Transactions on Neural Networks and Learning Systems*, 30, 11, 2019.

[10] U. Batool, M. I. Shapiai, M. Tahir, Z. H. Ismail, N. J. Zakaria, A. Elfakharany, "A Systematic Review of Deep Learning for Silicon Wafer Defect Recognition", *IEEE Access*, 9, 116573, 2021.

6

Failure Detection in Silicon Package

Saad Al-Baddai and Jan Papadoudis

Infineon Technologies AG, Germany

Abstract

In an ever more connected world, semiconductor devices represent the core of every technically sophisticated system. The desired quality and effectiveness of such a system through assembly and packaging processes is high demanding. In order to achieve an expected quality, the output of each process must be inspected either manually or rule-based. The latter would lead to high over-reject rates which require a lot of additional manual effort. Moreover, such an inspection is sort of handcrafted by engineers, who can only extract shallow features. As a result, either more yield-losses due to an increase in the rejection rate or more products with low quality will be shipped. Therefore, the demand for advanced image inspection techniques is constantly increasing. Recently, machine learning and deep learning algorithms are playing an increasingly critical role to fulfil this demand and therefore have been introduced in multiple applications. In this paper, an overview of the potential use of advanced machine learning techniques is explored by showcasing of image and wirebonding inspection in semiconductor manufacturing. The results are very promising and show that AI models can find failures accurately in a complex environment.

Keywords: anomaly detection, labelling, manufacturing AI solutions, AI integration, transfer learning, scalability.

DOI: 10.1201/9781003377382-6

6.1 Introduction and Background

Semiconductor manufacturing produces the most highly advanced microchips in the world. A manufacturing process of these chips goes through multiple sequences and interacting sub-processes and during that operates in extreme quality-demanding conditions. Thus, it has an increasing complexity and demand on quality requirements, as electronics increasingly become an important part of modern society. In principle semiconductor manufacturing is equipped with lots of sensors to monitor the processes but it lacks a suitable way to make use of this data. However, due to the complexity of the processes and unknown correlation among the collected data, such traditional techniques become quite limited. Here's where AI takes the initiative and offers a promising solution for feature extraction, condition monitoring and fault modelling for anomaly/defect detection using sophisticated algorithms [5]. Therefore, one of the success factors in optimizing the industrial processes is either automatic anomaly detection, supervised learning or both, which leads to prevent production flaws, herewith improving quality, increasing yields and making benefits. The popular way of anomaly detection in many of industrial application is by adjusting digital camera parameters or sensors during collecting images or time series data. This is basically an image or signal anomaly detection problem that is searching later on for patterns that are different from normal data [4]. As a human one can easily manage such task by recognizing of normal patterns, but this is relatively not easy for machines. Unlike other classical approach, image anomaly detection faces some of the following difficult challenges: class imbalance, quality of data, and unknown anomaly [4]. A prevalence of abnormal events are generally exception, whereas normal events account for a significant proportion. Some techniques usually handle the anomaly detection problem as a "one-class" problem. Here models learn by using the normal data as truth ground and afterwards evaluates whether the new data belong to this truth ground or not, by the degree of similarity to the truth ground. In the early applications of surface defect detection, the background is often modeled by designing handmade features on defect-free data. For example, Bennatnoun et al. used blobs technique [3] to characterize the correct texture and to detect deviations through changes in the charter ships of generated blobs. Amet et al. [2] used wavelet filters to extract different scales of defect-free images, then extracted the informative features of different frequency scales of images. However, most of these methods focus can work with homogeneous date with good quality and would require a prior knowledge. Generally, still some challenges which strongly depend on the field of application. Thus, there is no universal pattern or system, which does not directly allow to use techniques developed for one application to another. Thus, machine/deep learning offers promising solutions in such complex environment. However, the former can be adapted or scaled to other application or use cases. Due to these above-mentioned challenges unsupervised anomaly detection on multi-dimensional data is very highly demanding in machine learning and business applications [6]. Please note, this paper is extended of the published work in [1]. The latter focused on the data preparation, labelling techniques and preliminary results. A new contribution related to quantities, framework and transfer learning and scalability is presented. Therefore, a short description about the data is introduced. Then, labelling approach is shortly discussed. Afterwards, framework is depicted and effective of transfer learning is discussed. Finally, the results are showed and conclusion is drawn.

6.2 Dataset Description

This manuscript showcases dealing with time series data as well as with images at different processes during packaging. The data for the first case is collected in the early phase, at wirebonding process. These data are collected from three different sensors. Namely a current sensor, located at the transducer, a displacement sensor measuring the deformation of the wire respectively the path of the bonding tool and a frequency sensor, also located at the transducer of the wirebonder. Each of these sensors collects roughly 432 features during 143 timestamps. However, the collected data are highly redundant (see Figure 6.1). This is because there is multiple bond connection on one device which share the same process parameters and behave quite similar. However, sometimes, contamination of the device or a misadjusted machine would cause misaligned or deformed bonds, see Figure 6.1. Here, there is a need to develop a ML solution for detecting such deviations. Similarly, the biggest challenge of the outgoing optical inspection (OOI), in the second use case, is the defect detection on the heatsink, see Figure 6.1, which consists of a rough copper surface. It needs to inspected for scratches, metal or mold particles as well as for mechanical damage like imprints. However, this surface shows a very high variety in appearance, as it is oxidized during preceding high temperature testing steps. Hence, the inspection cannot be carried out using rule-based algorithms, as the oxidized areas cannot be distinguished clearly from true defects by a rule-based algorithm. In this

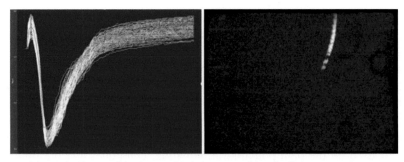

Figure 6.1 *Left*: Curve with abnormal minimum position (red) in comparison to normal ones (white) of recorded sensor data during wirebonding process. *Right*: shows an example of abnormal OOI image with shown crack on the surface.

context, trained personnel took care of the heatsink inspection and was used to label the image data, roughly 300 images, for supervised learning.

6.2.1 Data Collection and Labelling

Data labelling is an essential step in a machine learning area. Here, the common phrase "Garbage in - Garbage out" is used very commonly in the ML community, that means the quality of the model strongly depends on the quality of the (labelled) training data. In this work, two approached are considered:

- $X \rightarrow Y$

Indeed, data labelling is a task that requires a lot of manual work. In this approach, labelling data(images) is done based on human experience. Luckily, only few percent of data had to reviewed after applying the tool introduced in [1] for reducing the effort. This process is done by review the sorted data of historical images and recognize on the defects by looking closely at heat sink surface. Thus, there is no need for prior knowledge about the status of Y machine to sort out X data. Afterwards, simply, the data can be categorized into two categories as either healthy(good) or unhealthy(fail). These data, then, can be used for training the AI model. This approach is used for labelling the first case OOI.

- $Y \rightarrow X$

Contrary to the first approach, in this approach human's experience unfortunately is not fully helpful for labelling data, as the data is very complex. Hence, the design of experiment (DOE) is set by checking the machine status while collecting data. Therefore, a predefined mis-adjustment in Y wire bond should be known to get deviation on X data.

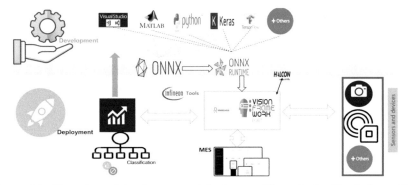

Figure 6.2 Flow chart of development and deployment life cycle for AI solution at IFX. In development phase data scientists could use different programming language as the final model can be converted to ONNX. In deployment phase, the vision frame can simply access to ONNX and run during inference time.

6.3 Development and Deployment

In order to satisfy the robustness requirements of AI model, we propose the AI framework to be adapted to the best practices with the following characteristics

- Short adaption cycles.
- Testing in every stage and automatically integration and deployment.
- Reproducible processes and reliable software releases.

Figure 6.2 shows a typical DevOps process which is the basis for continuous integration and delivery. Thus, the following feedback loops are added to the process in order to integrate central ML lifecycle steps:

- Define and build a suitable model and improve it based on demo feedback through experiments using any suitable programming language.
- Converting the optimal model, based on observed model performance, into ONNX (or other suitable format) and integrating it to the target AI platform.
- Retrain, when it is needed, an operational model based on new real-life data and report the performance.
- Adapt result of the whole process based on the performance of the models on productive data.

However, for deployment, it gets more complex, because of additional types of IFX infrastructure must be considered. Here, Figure 6.3 shows the process which is extended by the new development into the existing IFX

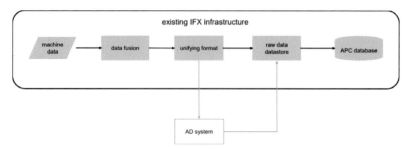

Figure 6.3   Process flow integration of the developed AD solution into an existing IFX infrastructure.

infrastructure. From the perspective of a classic ML lifecycle, the role setting of Business Analysts together with Data Scientists and Data Engineers is sufficient for conducting a working ML solution which proves to deliver all required benefits.

6.4 Transfer Learning and Scalability

Transfer learning is simply fine-tuning previously trained neural networks. In this context we transfer the trained model on OOI data into other processes of packaging, see Figure 6.4 Thus, instead of creating an AI model from scratch, only a few images of the new process are enough for fine tuning the pre-trained model of OOI images. Interestingly, not only the collected images from new process are similar to the OOI images but the defect types as well. As a result, the model reports a high accuracy as is shown in Table **??**. The anomaly detection for the wire bonding process has a wide range of application, as there are multiple Infineon sites and multiple machines of the same type. The training of an anomaly detection model can benefit from unlabelled data under the assumption that the majority of the data is good. Given the general high yield this assumption is valid. Given multiple similar machines there are two approaches to scale one model to multiple machines.

- Using data from multiple machines for the training. Thus, the model implicitly learns differences between the machines and the same model can be used for multiple machines.
- Using an anomaly detection model, which was trained on a prior defined machine and setting up all other machines to behave most similar to the selected machine. Thus, all other machines generate raw data of the same input space as the selected machine.

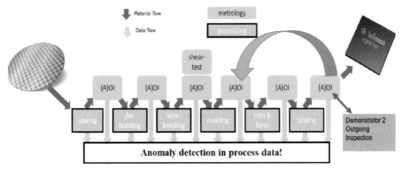

Figure 6.4 show the flow processes during silicon package, the backside blue arrow shows the position of transfer learning from OOI backwards to taken images after molding process, see Figure 6.5

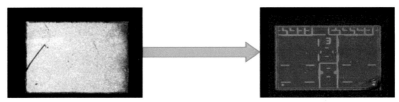

Figure 6.5 shows an example of the OOI image on left side (This image is taken before shopping and after electrical test) and example of image after molding process on right side.

With this procedure it was possible to scale one model to s complete production line with more than 30 machines.

6.5 Result and Discussion

For wire bonding use case, two different approaches to validate the system were made. The first one was to simply calculate the percentage of devices which showed an anomaly in the dataset and compare this to the process yield. If these percentages align this is a good indicator that the anomaly detection represents the product quality. Additionally, a statistical significantly correlation between high anomaly values and bad electrical test results is considered. For the second approach, we gathered multiple devices which showed a high anomaly value and examined them thoroughly. In all of the cases different influences could be found on the device, like a contaminated device, reduced shear value or input material which was out of specifications. But not all findings, even though varying from the normal, will lead to a malfunctioning device. However, an important aspect of the used anomaly detection was that the result is an anomaly score, indicating how different the raw data from normal is not a Boolean indication anomaly / no anomaly. Thus, it is necessary to find an optimal threshold on which the difference in the raw data influences the quality of the product. An important impact of the work was also the adaptation of the approach to a performant data management infrastructure; i. e. the development of automatable methods for the detection of conspicuous parameter behaviour and its marking and storage. The evaluation was based on sample data and statistical analysis of standard deviations considering Nelson's rules. The work carried out covers both the familiarization with the various technologies and their variants, the adaptation of the methods to the subject area, and the prototypical implementation and testing of the algorithms by embedding them in automated analysis pipelines. Currently the anomaly detection for wirebonding is running on over 40 machines on 3 different IFX sites. During a runtime of 4 months, several misadjusted bonders were detected, random errors and contaminated devices. However, currently a big focus is set to fully integrate the model not only in the infrastructure but also in the day to day workflow of the operators, this also includes a clear definition of action plans for found deviations and trainings of operators. For OOI use case, after collecting images, the labeled images are pre-processed first by cropping the region of interest and normalization the intensity values between 0 and 1. These images are sent to CNN for training purpose. The CNN consist of 100 layers. The latter consisting of different blocks. Each block contains the convolutional, pooling and ReLU layer. Also, before the last layer, fully connected layer, a strict regularization factor is added in order to avoid over-fitting issue by adding dropout layer with value 0.6. The data was splited into 80% training and 20% validation data. The model reported with accuracy higher than 99%. Afterwards, the model is tested on productive data with roughly 25k images. Table 6.1 shows the confusion matrix with the important measures, sensitivity, specificity and accuracy. As, one can see that model to follow zero defect philosophy, as sensitivity value is 100%. The accuracy also is less than 1%. Hence, only the latter have to be reviewed by an expert. Moreover, the performance model after scaling to anew process is still very robust. As one can see in the Table 6.2, which shows the reported results by a model when run on productive data of the new process. Although, one can see there is one escapee in bottom surface (BOT), but the accuracy is still higher than 99%.

Table 6.1 Show the confusion matrix and metrics of the CNN model on productive data for BOT and TOP of OOI images.

BOT		Defect	Good		TOP		Defect	Good
	Defect	250	379			Defect	130	220
	Good	0	39921			Good	0	25000
	Acc.:99,07%	Sen.:100%	Spe.:99,06%			Acc.:99,13%	Sen.:100%	Spe.:99,13%

Table 6.2 Show the confusion matrix and metrics of the CNN model on productive data for BOT and TOP of the new process.

BOT		Defect	Good		TOP		Defect	Good
	Defect	227	198			Defect	751	60
	Good	0	26063			Good	1	9353
	Acc.:99,25%	Sen.:100%	Spe.:99,25%			Acc.:99,40%	Sen.:99,87%	Spe.:99,36%

6.6 Conclusion and Outlooks

In this paper, two use cases show the potential benefits of using AI models in detecting abnormalities in industrial packages. Moreover, the methodology shows the possibility of scaling such solutions to new similar use cases or machines with minimum effort. As a result, not only the manual effort would significantly be reduced, but also costs and the quality of the products would be improved. Additionally, the long-term goal is not only to find the deviation but to detect exactly the root cause behind it. However, there is still a lot of work left, unrealized potentials benefit of AI solutions, but IFX has already taken a step forward in the right direction. Thus, semiconductor community is investing more with AI to harvest its benefits in the short and, most importantly, long term. Generally, the results are promising and would be a good alternative to classical approaches. The next steps are monitoring, optimization and more validation for both solutions in a productive environment.

Acknowledgements

AI4DI receives funding within the Electronic Components and Systems for European Leadership Joint Undertaking (ECSEL JU) in collaboration with the European Union's Horizon2020 Framework Programme and National Authorities, under grant agreement n° 826060.

References

[1] S. Al-Baddai, M. Juhrisch, J. Papadoudis, A. Renner, L. Bernhard, C. Luca, F. Haas, and W. Schober. Automated Anomaly Detection through Assembly and Packaging Process, pages 161–176. 09 2021.

[2] A. Amet, A. Ertuzun, and A. Ercil. Texture defect detection using subband domain co-occurrence matrices. pages 205 – 210, 05 1998.

[3] A. Bodnarova, M. Bennamoun, and K. Kubik. Automatic visual inspection and flaw detection in textile materials: A review. pages 194–197, 01 2001.

[4] T. Ehret, A. Davy, J. M. Morel, and M. Delbracio. Image anomalies: a review and synthesis of detection methods. 08 2018.

[5] G. A. Susto, M. Terzi, and A. Beghi. Anomaly detection approaches for semiconductor manufacturing. Procedia Manufacturing, 11:2018–2024, 12 2017.

[6] B. Zong, Q. Song, M. R. Min, W. Cheng, C. Lumezanu, D. Cho, and H. Chen. Deep autoencoding gaussian mixture model for unsupervised anomaly detection, 2018.

7

S2ORC-SemiCause: Annotating and Analysing Causality in the Semiconductor Domain

Xing Lan Liu[1], Eileen Salhofer[1,2], Anna Safont Andreu[3,4], and Roman Kern[2]

[1]Know-Center GmbH, Austria
[2]Graz University of Technology, Austria
[3]University of Klagenfurt, Austria
[4]Infineon Technologies Austria

Abstract
For semiconductor manufacturing, easy access to causal knowledge documented in free texts facilitates timely Failure Modes and Effects Analysis (FMEA), which plays an important role to reduce failures and to decrease production cost. Causal relation extraction is the tasks of identifying causal knowledge in natural text and to provide a higher level of structure. However, the lack of publicly available benchmark causality datasets remains a bottleneck in the semiconductor domain. This work addresses this issue and presents the S2ORC-SemiCause benchmark dataset. It is based on the S2ORC corpus, which has been filtered for literature on semiconductor research, and consecutively annotated by humans for causal relations. The resulting dataset differs from existing causality datasets of other domain in the long spans of causes and effects, as well as causal cue phrases exclusive to the domain semiconductor research. As a consequence, this novel datasets poses challenges even for state-of-the-art token classification models such as S2ORC-SciBERT. Thus this dataset serves as benchmark for causal relation extraction for the semiconductor domain.

Keywords: causality, relation extraction, information extraction, bertology, annotation.

91

DOI: 10.1201/9781003377382-7

7.1 Introduction

Although causality represents a simple logical idea, it becomes a complex phenomenon when appearing in textual form. Natural language provides a wide variety of structures to represent causal relationships that can obfuscate the causal relations expressed via cause and effect. The task of causal relation extraction aims at extracting sentences containing causal language and identifying causal constituents and their relations [17].

In the last years significant progress have been made in automatizing the identification of causal cues and extraction of causal relation in natural language, defining it as a multi-way classification problem of semantic relationships [6], designing a lexicon of causal constructions [2, 3], and insights how to achieve high inter-rater agreement [13]. Approaches have been developed in scientific domains traditionally dominated by textual information, such as biomedical sciences. Here, models to process causal relations are facilitated and accelerated with the development of benchmark datasets such as BioCause [10]. Such datasets not only allow for comparison and automatic evaluation of custom causal extractors, but also allow for training high performing supervised models.

For semiconductor manufacturing, much of existing knowledge can be considered to be causal, highlighted by approaches like Ishikawa causal diagrams as well as the Failure Modes and Effects Analysis (FMEA) tool which captures root causes of potential failures. Even though such FMEA document provides more structure than natural language text, dedicated pre-processing is required before further processing [12]. A signification amount of such causal knowledge is captured in textual documents, such as reports and knowledge bases. However, there is no publicly available annotated dataset for causal relation extraction yet. As a consequence, in this work we propose such a dataset, named *S2ORC-SemiCause*. The source for the documents of this novel dataset is the S2ORC academic corpus, which has been filtered for documents of relevance for the semiconductor domain. Human annotators identified causal cues and causal relations in the documents of the corpus. To achieve consistent and reproducible results, an annotation guideline was created and the annotation processes was conducted in multiple phases. To provide baseline performance, the pre-trained language model BERT [1], which is currently considered state of the art for many natural language processing (NLP) tasks was adapted for the task. An error analysis gives insights on the challenges of future causal relation extraction methods.

In summary, our main contributions are:

- *S2ORC-SemiCause*, a causality dataset for the semiconductor domain that aims to provide a benchmark for causal relation extraction performances and facilitate research on dedicated methods;
- Practical annotation guidelines designed to yield high inter-annotator agreement for semiconductor literature, to enable the creation of further, similar datasets;
- Identified the key differences of *S2ORC-SemiCause* compared to other domains, and highlighted the resulting challenges for state-of-the-art NLP models.

7.2 Dataset Creation

7.2.1 Corpus

Our semiconductor corpus is selected from the 24 million papers in the engineering and related domains from the S2ORC corpus [8] (total 81.8 million papers). The subdomain is further filtered using a series of keywords specific for the semiconductor domain, such as device locations, electrical and physical faults, technologies (e.g. SFET), Focused Ion Beam, etc. For a paper to be selected, it needs to include at least four of these keywords.

From the resulting subset of 21 thousand papers, 400 abstract and 400 paragraphs are randomly sampled, among which 600 sentences are selected randomly for annotation.

7.2.2 Annotation Guideline

We have adapted the annotation guidelines[1] from the creation of BECauSE Corpus 2.0 [3]. The main differences are (1) the relation types "Motivation" and "Purpose" are further merged into one type (name "Purpose") since it is found from previous work [5] that annotators have difficulty distinguishing these two types; (2) *"max-span" rule*, namely, the span should include full phrase or clause. The *"max-span"* rule not only retains important context information for the causal relations, but also enables higher inter-annotator agreement. This was also motivated that it assumed to be easier to automatically reduce a phrase to its heads, instead of expanding a short, existing annotation.

---

[1]The annotation guideline will be make public at https://github.com/tugraz-isds/kd.

Table 7.1   Inter-annotator agreement for the first two iterations. *Arg1* (cause) refers to the span of the arguments that lead to *Arg2* (effect) for the respective relation type.

	Iteration 1	Iteration 2
Relation classification Cohen's κ	0.65	0.80
Consequence Arg1 F_1	0.55	0.71
Consequence Arg2 F_1	0.60	0.81
Purpose Arg1 F_1	0.00	0.92
Purpose Arg2 F_1	0.00	0.80
F_1 micro average	0.49	0.78

Table 7.2   Comparison of labels generated by both annotators for Iteration 2. Examples and total counts (in number of arguments) for each type also given. Arg1 and Arg2 are highlighted with blue and yellow background, respectively. Partial overlapped texts are highlighted with green background.

Type	#	Example sentence
Exact match	54	*In fact, and for the soil in question,* the capillary rise process is low *, so the indirectly loss by evaporative loss is low too* .
Partial overlap	8	*This result suggests a possible dynamical influence of* the mesospheric layers *on* the lower atmospheric levels .
Only one annotator	14	*The wing displaces away from the ground , as a result of* the reduction in (-ve) lift .

7.2.3 Annotation Methodology

Since the annotations should contain as little ambiguity as possible, we aimed to design a methodology to achieve consistent annotations. To this end, the dataset was annotated in a total of 3 iterations. For the first two iterations with 50 sentences each, both annotators label the same set, so that inter-annotator-agreement (IAA) can be evaluated. Between the two iterations, the two annotators discussed the results and updated the guideline.

Table 7.1 shows that there are significant improvement in Inter-Annotator Agreement (IAA) from iteration 1 to iteration 2, both in terms of Cohen's κ, and F_1. The main improvement comes from (1) direction for *Purpose* relation (namely, *arg2* should be the purpose); (2) *"max-span" rule*, namely, the span should include full phrase or clause.

With Iteration 2, the two annotators reached a substantial agreement, where both Cohen's κ for relation classification and F_1 for argument spans are around 0.8. For reference, in Dunietz et al. [3] a Cohen's κ of 0.70 was reported for the relation type. Results of detailed inspection are summarized

Table 7.3 Descriptive statistics of benchmark datasets. Overview of CoNLL-2003 (training split) and BC5CDR (training split) for named entity recognition, as well as causality dataset BioCause (full dataset), and S2ORC-SemiCause (training split).

	CoNLL-2003	**BC5CDR**	**BioCause**	**S2ORC-SemiCause**
#sentences	14,042	4,612	37,422	360
Avg. sentence length (in tokens)	14.5	25.0	7.8	32.0
Avg. argument length (in tokens)	1.4	1.5	3.6	9.5

in Table 7.2. For 54 arguments, both annotators agree in both span and argument type. The remaining disagreements are from (1) one annotator misses a relation (14 occurrences); (2) only partial overlap of the annotated spans by both annotators (8 occurrences).

Based on the insights from the updated baseline, the first set of document was revisited and both set of annotations from the first two iterations were then merged manually. In addition, for the 3rd iteration, two extra sets of 250 sentences were annotated by each annotators. As a result, our dataset consist of 600 sentences annotated with Consequence and Purpose relations.

7.2.4 Dataset Statistics

We notice that compared to other benchmark NER datasets, such as CoNLL2003 [4], BC5CDR [7], and BioCause [10] (see Table 7.3), S2ORC-SemiCause dataset differs in terms of (1) smaller size; (2) longer sentence length; (3) longer argument length. While data size is found to be generally sufficient for entity recognition tasks [14], and longer sentence length is found to be preferred [14], the effect of longer argument length remains to be evaluated.

7.2.5 Causal Cue Phrases

When present, the causal cue phrases are also annotated. Figure 7.1 depicts the most common cue phrases for both relation types. *"To"* is the most frequently occurring cue because it is by far the most dominating cue phrase for relation type *purpose*. The cue phrases for *consequence* are much more diverse. Compared to other corpus of general domain [9, 11], in S2ORC-SemiCause dataset, cue words such as *increase, decrease, improve, reduce* are also ranked very high.

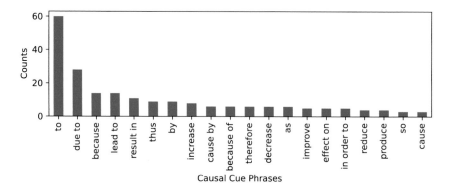

Figure 7.1 Causal cue phrases ranked by frequency for all sentences in S2ORC-SemiCause dataset.

7.3 Baseline Performance

To establish a point of reference for the community, we provide an initial baseline performance. For the baseline approach we considered the causal relation extraction task as an sequence classification task. As a technical realisation, we fine-tuned BERT on the down-stream task of token-level classification [1]. An error analysis is then performed to identify the main challenges in extracting causal relations from scientific publications in semiconductor research.

7.3.1 Train-Test Split

The total 600 sentences are split into training, validation, and test sets, with the ratio $60 : 20 : 20$, stratified on relation type[2]. In addition, also the iterations were stratified evenly to avoid unwanted biases. The descriptive statistics for each split is listed in Table 7.4.

7.3.2 Causal Argument Extraction

As recommended in [1], which describes a similar scenario, we considered the task as a token-level classification. Namely, a pretrained BERT model is stacked with a linear layer on top of the hidden-states output, before fine-tuned on training examples. And the pretrained S2ORC-SciBERT model [8] is selected for fine-tuning using transformers library from Hugging Face [16].

---

[2]We release all data for future studies at https://github.com/tugraz-isds/kd

Table 7.4 Descriptive statistics of *S2ORC-SemiCause* dataset. *#-sent*: total number of annotated sentences, *#-sent no relations*: number of sentences without causality, *Argument*: total amount and mean length (token span) of all annotated argument, *Consequence/Purpose*: amount and mean length of cause and effect arguments for the respective relation types.

	#-sent	#-sent no relations	Argument		Consequence				Purpose			
					cause		effect		cause		effect	
			count	mean	count	mean	count	mean	count	mean	count	mean
overall	600	291	670	9.4	258	8.4	290	9.2	58	10.8	64	12.9
train	360	174	405	9.5	155	8.5	178	9.1	34	11.1	38	13.5
dev	120	55	122	9.3	49	8.1	52	8.8	10	9.7	11	16.1
test	120	62	143	9.3	54	8.3	60	9.9	14	10.9	15	8.9

Table 7.5 Baseline performance using BERT with a token classification head. Both the F_1 scores and the standard derivation over 7 different runs are shown. Despite the small sample size, the standard deviation remain low, similar to previous work [14].

Relation	Argument	#	F_1	F_1-filter	F_1-filter partial
Consequence	Arg1	54	0.43 ± 0.03	0.48 ± 0.02	0.59 ± 0.01
Consequence	Arg2	60	0.45 ± 0.03	0.50 ± 0.03	0.62 ± 0.02
Purpose	Arg1	14	0.20 ± 0.07	0.25 ± 0.10	0.50 ± 0.05
Purpose	Arg2	15	0.31 ± 0.06	0.36 ± 0.08	0.57 ± 0.07
	micro average	143	0.39 ± 0.02	0.45 ± 0.02	0.59 ± 0.01

The resulting F_1 scores[3] are shown in Table 7.5 and is remarkable lower than for other benchmark NER datasets when down-sampled to similar size [14].

7.3.3 Error Analysis

In order to understand the causes for the low F_1 score of the baseline model, an error analysis is performed.

Length of Argument Span

Firstly, a manual inspection revealed that for 30 ± 4 (out of the total 120) sentences, the fine-tuned model predicts sequences similar to $[O\ I\ I\ \cdots\ ]$, i.e., the models did not learn that an argument must always start with a "B" type with the IOB (Inside–Outside–Beginning) notation.

We hypothesize that this might be because our argument spans are much longer than other datasets (see Table 7.4 and Table 7.3). As a result, either the self-attention might no longer efficiently keep track of the $[B\ I\ \cdots\ ]$ pattern, or the over-abundant "I" class might bias the model loss.

---

[3]The best performance is found using learning rate $1.5e - 4$, batch size 8, warm up steps 10, and 10 epochs.

Table 7.6 Comparison of predicted and annotated argument spans for the test split. Examples and total counts (in number of arguments) for correct prediction and for each error source are also given. Arg 1 and Arg 2 are highlighted with blue and yellow background, respectively. Partial overlapped texts are highlighted with green background.

Type	#	Example sentence
Exact match	68	*The broad peak at 5 eV is due to N(2p) electrons* .
Partial overlap	41	*These safe zones are provided to a model predictive controller as reference to generate feasible trajectories for a vehicle* .
Spurious	46	*The roles of initial concentrations , space dimension and ratio of the reactant diKusinties in the modification of the reaction rate by many - particle eMects are compared with computer simulations.*
Missed	34	*This result validates the bolometric IR luminosities derived from MIR luminosities* .

Following this hypothesis, we expect better performances for shorter arguments than for longer. Indeed we observe that correct predictions are shorter by 2.7 tokens on average ($p\_value = 0.008$).

To quantify the effect of such incorrect $[O\ I\ I\ \cdots]$ sequences, we re-evaluated F_1 score after filtering out such predictions. The results are shown in Table 7.5 as "F_1-filter", and an improvement of 6 points is observed compared to the F_1 score before filtering.

Predictions with Partial Overlap

Out of the predicted argument, 41 were counted as incorrect, but overlapped partially (see example in Table 7.6), and manual inspection suggest that they often contain valid causal information.

Following [15], the model performance can be evaluated taking into account partial overlaps. The results are listed in Table 7.5 as "F1-filter partial", and the average F1 score becomes 0.59, which is about 80% of human performance (inter-annotator F1 value of 0.78), and is inline with the sample-size scaling as reported previously [14].

Spurious and Missed Predictions

Spurious examples (false positives) are the cases where the model predicts a relation while annotators do not label. After manual inspection, we find it arguable that some *spurious* predictions made by the model might actually be valid causal relations as well. For example, the spurious example shown in Table 7.6 is arguably causal as well following the (*The role of ... in ...*) construct.

Missed examples (false negatives) are the cases where annotators have labelled while the model fails to predict a relation. For example, the missed example shown in Table 7.6 uses the rare causal trigger *derived from*, which might be the reason why the model failed to recognize.

7.4 Conclusions

Causality is critical knowledge in semiconductor manufacturing. In order to enable automatic causality recognition, we created the *S2ORC-SemiCause* dataset by annotating 600 sentences with 670 arguments for causal relation extraction from a subset of semiconductor literature taken from the S2ORC dataset. This unique dataset challenges established state-of-the-art techniques, because of its long spans for each argument. This benchmark dataset is intended to spur further research, fuel development of machine learning models, and to provide benefit to the NLP research in semiconductor domain.

Acknowledgements

The research was conducted under the framework of the ECSEL AI4DI "Artificial Intelligence for Digitising Industry" project. The project has received funding from the ECSEL Joint Undertaking (JU) under grant agreement No 826060. The Know-Center is funded within the Austrian COMET Program– Competence Centers for Excellent Technologies under the auspices of the Austrian Federal Ministry of Transport, Innovation and Technology, the Austrian Federal Ministry of Economy, Family and Youth and by the State of Styria. COMET is managed by the Austrian Research Promotion Agency FFG. We acknowledge useful comments and assistance from our colleagues at Know-Center and at Infineon.

References

[1] J. Devlin, M. W. Chang, K. Lee, and K. Toutanova. BERT: Pre-training of deep bidirectional transformers for language understanding. *NAACL HLT 2019 - 2019 Conference of the North American Chapter of the Association for Computational Linguistics: Human Language Technologies - Proceedings of the Conference*, 1(Mlm):4171–4186, 2019.

[2] J. Dunietz, L. Levin, and J. Carbonell. Annotating causal language using corpus lexicography of constructions. *The 9th Linguistic Annotation Workshop held in conjunction with NAACL 2015*, (2014):188–196, 2015.

[3] J. Dunietz, L. Levin, and J. G Carbonell. The because corpus 2.0: Annotating causality and overlapping relations. In *Proceedings of the 11th Linguistic Annotation Workshop*, pages 95–104, 2017.

[4] E. F. Tjong Kim Sang, and F. De Meulder. Introduction to the CoNLL-2003 shared task: Language-independent named entity recognition. In *Proceedings of the Seventh Conference on Natural Language Learning at HLT-NAACL 2003*, pages 142–147, 2003.

[5] D. Gaerber. Causal information extraction from historical german texts, 2022.

[6] I. Hendrickx, S. N. Kim, Z. Kozareva, P. Nakov, D. Ó Séaghdha, S. Padó, M. Pennacchiotti, L. Romano, and S. Szpakowicz. SemEval-2010 task 8: Multi-way classification of semantic relations between pairs of nominals. In *Proc. of the 5th Int. Workshop on Semantic Evaluation*, pages 33–38, Uppsala, Sweden, 2010. Association for Computational Linguistics.

[7] J. Li, Y. Sun, R. J. Johnson, D. Sciaky, C.-H. Wei, R. Leaman, A. P. Davis, C. J. Mattingly, T. C. Wiegers, and Z. Lu. Biocreative V CDR task corpus: a resource for chemical disease relation extraction. *Database*, 2016.

[8] K. Lo, L. L. Wang, M. Neumann, R. Kinney, and D. Weld. S2ORC: The semantic scholar open research corpus. In *Proceedings of the 58th Annual Meeting of the Association for Computational Linguistics*, pages 4969–4983, Online, July 2020. Association for Computational Linguistics.

[9] Z. Luo, Y. Sha, K. Q. Zhu, S. W. Hwang, and Z. Wang. Commonsense causal reasoning between short texts. *Proc. Int. Workshop Tempor. Represent. Reason.*, pages 421–430, 2016.

[10] C. Mihăilă, T. Ohta, S. Pyysalo, and S. Ananiadou. BioCause: Annotating and analysing causality in the biomedical domain. *BMC Bioinformatics*, 14, 2013.

[11] S. Pawar, R. More, G. K. Palshikar, P. Bhattacharyya, and V. Varma. Knowledge-based Extraction of Cause-Effect Relations from Biomedical Text. 2021.

[12] H. Razouk and R. Kern. Improving the consistency of the failure mode effect analysis (fmea) documents in semiconductor manufacturing. *Applied Sciences*, 12(4), 2022.

[13] I. Rehbein and J. Ruppenhofer. A new resource for German causal language. In *Proceedings of the 12th Language Resources and Evaluation Conference*, pages 5968–5977, Marseille, France, May 2020. European Language Resources Association.

[14] E. Salhofer, X. L. Liu, and R. Kern. Impact of training instance selection on domain-specific entity extraction using bert. In *NAACL SRW*, 2022.

[15] I. Segura-Bedmar, P. Martínez, and M. Herrero-Zazo. SemEval-2013 task 9 : Extraction of drug-drug interactions from biomedical texts (DDIExtraction 2013). In *Proc. of the 7th Int. Workshop on Semantic Evaluation (SemEval 2013)*, 2013.

[16] T. Wolf, L. Debut, V. Sanh, J. Chaumond, C. Delangue, A. Moi, P. Cistac, T. Rault, R. Louf, M. Funtowicz, J. Davison, S. Shleifer, P. v. Platen, C. Ma, Y. Jernite, J. Plu, C. Xu, T. L. Scao, S. Gugger, M. Drame, Q. Lhoest, and A. Rush. Transformers: State-of-the-art natural language processing. In *Proc. of the 2020 Conf. on Empirical Methods in NLP: System Demonstrations*, 2020.

[17] J. Yang, S. C. Han, and J. Poon. A survey on extraction of causal relations from natural language text. *Knowledge and Information Systems*, pages 1–26, 2022.

8

Feasibility of Wafer Exchange for European Edge AI Pilot Lines

Annika Franziska Wandesleben[1]*, Delphine Truffier-Boutry[2]*,
Varvara Brackmann[1], Benjamin Lilienthal-Uhlig[1], Manoj Jaysnkar[3],
Stephan Beckx[3], Ivan Madarevic[3], Audde Demarest[2], Bernd Hintze[4],
Franck Hochschulz[5], Yannick Le Tiec[2], Alessio Spessot[3],
and Fabrice Nemouchi[2]

[1]Fraunhofer IPMS CNT, Germany
[2]Université Grenoble Alpes, CEA-Leti, France
[3]imec, Belgium
[4]FMD, Germany
[5]Fraunhofer IMS, Germany
*Equal contribution

Abstract

This paper compares the contamination monitoring of the three largest microelectronics research organizations in Europe, CEA-Leti, imec and Fraunhofer. The aim is to align the semiconductor infrastructure of the three research institutes to accelerate the supply to European industry for disruptive chip processing. To offer advanced edge AI systems with novel non-volatile memory components, integration into state-of-the-art semiconductor fabrication production flow must be validated. For this, the contamination monitoring is an essential aspect. Metallic impurities can have a major impact on expensive and complex microelectronic process flows. Knowing this, it is important to avoid contamination of process lines. In order to benefit from the combined infrastructure, expertise and individual competences, the feasibility of wafer loops needs to be investigated.

Through a technical comparison and a practical analysis of potential cross-contaminations, the correlation of the contamination measurement

DOI: 10.1201/9781003377382-8

results of the research institutes is investigated. The results demonstrate that the three institutes are able to analyse metallic contamination with comparable Lower Limits of Detection (LLDs). This result sets the foundations for smooth and fast wafer exchange for current and future needs, potentially not only within research institutes as well as with industrial and foundry partners. The present work pays attention to both surface and bevel contamination. The latter requires very specific contamination collection which was also compared. Nevertheless, some challenges need to be addressed in the future to advance and accurate contamination monitoring.

Keywords: contamination, contamination monitoring and management, TXRF, VPD-ICPMS, surface, bevel, wafer loops.

8.1 Introduction

The aim is to align the semiconductor infrastructure of the three largest microelectronics research institutions in Europe, CEA-Leti, imec and Fraunhofer, in order to accelerate supply to European industry for disruptive chip processing. Contamination monitoring is an essential aspect of this alignment. Metallic impurities can have a major impact on expensive and complex microelectronic process flows. Therefore, it is important to avoid contamination of the process lines. To benefit from the semiconductor infrastructure, expertise and individual skills, the feasibility of wafer loops needs to be investigated. Additionally, to offer advanced edge AI systems with novel non-volatile memory components, integration into state-of-the-art semiconductor fabrication production flow must be validated. Metallic contamination can have a major impact within the microelectronic process flow, whereby the different chemical elements have various effects. Therefore, contamination of the process lines must be avoided (Bigot, Danel, & Thevenin, 2005; Borde, Danel, Roche, Grouillet, & Veillerot, 2007). To simplify the future exchange of wafers in-between research institutes and between institutes and semiconductor fabs, it is necessary to find out more about contamination monitoring and possible cross-contamination. For this purpose, a technical comparison and a practical analysis of the possible cross-contaminations is carried out in order to furthermore investigate the correlation of the contamination measurement results of the three institutes.

Table 8.1 Contamination monitoring techniques LETI / IMEC / FhG

Technique	LETI	IMEC	FhG
VPD-ICPMS	Wafer surface analysis	Wafer surface analysis	Wafer surface analysis
	Back side	Back side	Back side
	Front side	Front side	Front side
	Bevel	Bevel	*Bevel under development*
TXRF	Wafer surface analysis	Wafer surface analysis	*For wafer fragments and not yet available, under development*
	Back side	Back side	
	Front side	Front side	
	Bevel/Edge	Edge	

8.2 Technical Details and Comparison

The common techniques for contamination monitoring are TXRF and VPD-ICMPS. The three largest microelectronics research organizations in Europe, CEA-Leti, imec and Fraunhofer, are equipped with VPD-ICPMS while imec and CEA-Leti additionally use TXRF tools. The type of tool, its set up and qualification depend on the contamination management strategy developed in each clean room.

The capabilities of the individual institutes are summarised in the following Table 8.1.

8.2.1 Comparison TXRF and VPD-ICPMS Equipment for Surface Analysis

TXRF is ideal for high throughput applications as the measurements are based on the interaction of electron beams and silicon surfaces, without chemical manipulation. This technique allows to analyse fast enough both standard and noble elements in automatic mode with the possibility to localize the contamination on wafer with the mapping option. However lower limits of detection (LLD) are quite high, from $1E+9$ to $1E+11$ at/cm^2.

Concerning the VPD-ICPMS technique, it requires different chemical solutions for the collection of standard and noble elements, so campaigns need to be planned and there is no local resolution of contaminants. However, the collection of all metallic contaminants in a small droplet of chemistry induces significantly improved LLD values for all elements.

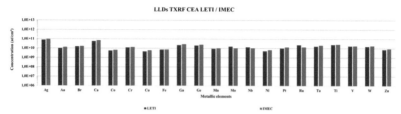

Figure 8.1   Comparison of TXRF LLDs of CEA LETI / IMEC

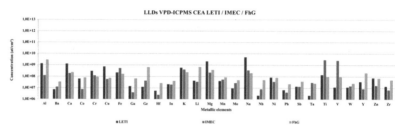

Figure 8.2   Comparison of VPD-ICPMS LLDs of CEA LETI / IMEC / FhG

To compare metallic contamination results obtained by the different institutes, the first goal was to compare LLDs of each element of each institute and
how it is experimentally determined. Indeed, LLD is the lowest concentration
at which an element can be reliably detected and is a key point for the
control of the metallic contamination at very low level. Depending on the
equipment, there are several ways to determine the LLD, and hence the need
for comparing the capabilities of each institute.

For TXRF, LLD values are nearly identical for each element, as shown
in Figure 8.1. As this technique is based on physical principles and since
both institutes have the same equipment (Rigaku TXRF), capabilities of both
institutes are the same. All LLDs are between 5E+9 and 5E+10 at/cm^2. Only
Ca and Ag are a little bit higher because Ca comes from the manual wafer
manipulation and Ag results from a high background noise on the TXRF
spectrum near 3 keV (Lα1 ray of Ag at 2.983 keV).

In case of VPD-ICPMS technique, the LLD results are not the same
across the three institutes. This can be explained by the fact that the technique
is based on chemical collection and each institute has its own specific system
with different approaches to the analysis and calculation of LLDs, as shown
in Table 8.2.

Table 8.2 Overview VPD-ICPMS LLD determination and technical details for LETI / IMEC / FhG

Aligned Data	LETI	IMEC	FhG IPMS CNT
Determination of LLD (VPD-ICPMS)	LLD VPD-ICPMS = 3xSigma for each elements	Calculated from 3xstandard deviation of calibration blank and slope of calibration curve.	For complete process VPD-ICPMS permanent blank method.
VPD Brand and type	Rigaku VPD300A, stand alone	IAS Expert^{TM} VPD system	**External source:** no data **CNT:** TePla System stand alone
ICP-MS brand and type	Agilent 8800, three quadrupoles	Perkin-Elmer Nexion^{TM} ICP-MS	**External source:** no data **CNT:** Thermo Fischer RQ, single quadrupole
Exclusion size VPD	7 mm	1 mm	**External source:** no data **CNT:** 5 mm (planned)

Figure 8.2 shows that the VPD-ICPMS LLDs of each institute are between 1E+6 and 5E+9 at/cm^2, more or less three decades lower than TXRF ones.

Differences observed across LLDs of each institute are due to the different techniques used and the different environments. The collection system at CEA-Leti is not full automatic and technicians have to transfer a small container containing the chemical droplet from the VPD to the ICPMS. This container has to be manually cleaned between collection and all these manual steps contribute to the increased Na, Mg and Ca levels of contamination. However, these specific LLDs are still lower than 1E+10 at/cm^2 and these elements are usually not critical for the microelectronic device performances. For imec, high values of Ti and V seem to be due to specific detector settings that favours minimal peak interference for Ti and V. For other elements, all imec LLDs are lower as they use a fully automatic tool without manual steps. Fraunhofer has a comparable system to CEA-Leti, but it is still in the method development process and the current analyses are done externally on an automated system.

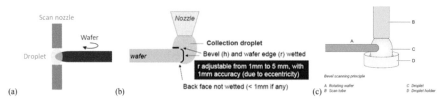

Figure 8.3   Schematic of the VPD bevel collection at (a) IMEC, (b) CEA-LETI and (c) FhG IPMS

Overall, the VPD-ICPMS LLDs of each institute are very low and comparable to industry standards and thus are sufficient for the metallic contamination control in the microelectronic environment. One other important parameter is the recovery rate that has to be more than 95 % for each of the elements. As each institute use the same chemical solution for the collection step, recovery rates are nearly the same and are very good (>95 %).

8.2.2 VPD-ICPMS Analyses on Bevel

For several years, wafer bevel contamination has become a challenge in the industry and it is therefore an increasing issue for R&D institutes. Actually, in order to increase device density on a wafer, individual chips need to be placed closer to the edge of a wafer limiting the waste of surface. In addition, wafers are increasingly processed by physical contact at the bevel, so this particular part of the wafer will need to be precisely controlled in the future. The full bevel area can only be analysed by VPD-ICPMS on bare Si wafers. Effectively, TXRF analysis of the full bevel is impossible because this technique is too sensitive to the topography and cannot quantify the metallic contamination localized on the fall of the bevel. The collection of contaminants at the bevel is a key point and each institute had to develop a specific system for the analysis. Thus, there are major technical variations between the collection systems used by the three institutes for the analysis of the bevel.

The Figure 8.3 shows the different techniques used by each institute for VPD collection on the bevel and the resulting different analysis surface. Therefore differences are also expected for the results of the VPD bevel analysis. Imec analyses the same area front side and back side 1 mm and the bevel, CEA-Leti analyses 5 mm front side, bevel and 1 mm back side. In Fraunhofer institute, the area is not defined yet as the method is still under

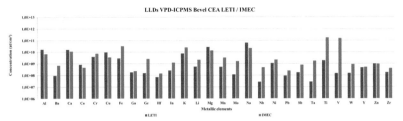

Figure 8.4 Comparison LLDs CEA LETI / IMEC for VPD-ICPMS Bevel

development. The monitoring of the bevel is another promising analytical technique and will be mandatory for the safe exchange of wafers, as with this control the probability of cross-contamination is further reduced.

Comparison of the LLDs for VPD-ICPMS bevel are shown in Figure 8.4. It shows that the LLDs are higher than those of the VPD-ICPMS surface since they are in the range of 1E+8 and 1E+11 at/cm^2. However, the values are quite similar and only Ti and V are noticeable again for imec due to their specific ICPMS detector setting.

8.3 Cross-Contamination Check-Investigation

In the frame of the present study, one equipment of each institute was selected for the control of the metallic contamination. Therefore, each institute chooses the tool that is regularly involved in the production memory flow and most critical in terms of contamination.

So called "witness wafers" were generated by each institute with the selected tool by handling bare Si wafers through the tool. In this way, the wafers are subjected to the specific tool contamination process. The analysis of the backside delivers information about the contamination by the handling system (chuck and robot). The analysis of the front side provides information about a possible contamination of the chamber. Afterwards, each institute characterises the metallic contamination of the wafers with their own techniques and finally, the analysis results are comprehensively evaluated.

8.3.1 Example for the Comparison of the Institutes

For the practical comparison of the measurement, the results of the three research institutes for a tool from Imec are presented as an example. The tool is a multi-module macro inspection, metrology and review tool for the front side of 200 mm and 300 mm wafers and additionally for the back side

Figure 8.5 Comparison TXRF results of CEA LETI / IMEC for IMEC inspection tool

and edge of 300 mm wafers. The tool supports the inspection of patterned and unpatterned wafers.

Figure 8.5 shows the comparison of TXRF measurement obtained by CEA-Leti and imec for the inspection tool. There is a high agreement between the values, demonstrating the comparability of the measurement results. The Ti measured by imec is assumed to be a handling contamination during the measurement. Nevertheless, the concentration is low.

Figure 8.6 shows the comparison of the VPD-ICPMS data for the back side surface of wafers. For the VPD-ICPMS, the results show noticeable differences. On Figure 8.6, only detected element at concentrations higher than the LLD are reported; i.e. if an element is not detected in one of institute, it is not mentioned in the graph. The first conclusion is that more elements are detected by VPD-ICPMS due to the lower LLDs. All the concentrations are lower than $1E+11$ at/cm^2 and are in accordance with TXRF results. The second conclusion is that the three analysed wafers have not the same contamination. If CEA-Leti and imec found Ga, Ge and Sb, Fraunhofer did not detect these elements. Imec and Fraunhofer quantified Al, Fe, Ti and W whereas CEA-Leti did not find these elements. The analysed wafers are not twins because the cross-contamination process do not allow to contaminate each wafers at the same concentration. Moreover, some wafers were more handled and shipped than other and these differences impact the metallic contamination.

Figure 8.7 shows the results obtained on the bevel. Contamination levels on the bevel are higher than those measured on the surface. In this example, results obtained by CEA-Leti and imec are in agreement when the elements are detected by both institutes. Concentrations measured by imec are almost higher than those of CEA-Leti, probably due to the different influencing factors. At first, collection techniques are different and the droplet scanned areas are not the same. Moreover, the bevel of each wafers was probably contaminated by the handling and the shipping. That is why concentrations

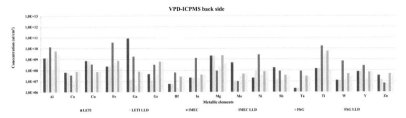

Figure 8.6 Comparison VPD-ICPMS results of CEA LETI / IMEC /FhG for IMEC inspection tool

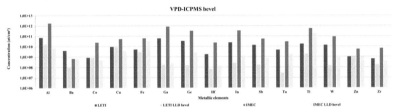

Figure 8.7 Comparison VPD-ICPMS bevel results of CEA LETI / IMEC for IMEC inspection tool

obtained on the bevel were always higher than those obtained on the surface. The study of the bevel is very challenging and these results show the metallic contamination due to the process in the selected equipment, but also those brought by the handling and the shipping.

8.4 Conclusiion

This study confirms that the three different institutes are able to analyse metallic contamination either by TXRF or VPD-ICPMS with comparable LLDs. This result is very promising for the exchange of wafers in the future. TXRF, with higher LLDs, did not show metallic contamination above 1E+11 at/cm^2. On the other side, due to very low limits of detection, VPD-ICPMS allows to observe different concentrations obtained by the different institutes. Nevertheless, these concentrations are very low. The cross-contamination in a tool do not allow to contaminate wafers at the same level. Hence in the future, in order to compare more reliably the capabilities of different institutes, an inter-laboratory test with intentionally standardised contaminated wafers would be necessary. Moreover, all the measurements were done on "witness wafers" and not on product-wafers. In the future, it will be necessary to develop techniques able to analyse the metallic contamination on real wafers during their flow. In this way, CEA-Leti has developed a new system allowing the metallic contamination control of the bevel of product wafers. (Boulard, et al., 2022) (FR Patentnr. U.S. Patent No 20200203190 A1, 2020).

Although some additional improvement is required to create a smooth loop between the research institutes, this work makes wafer exchange flow much easier due to the first experiences and contribute to the strengthening of the collaboration in current and future projects. Moreover, the conclusion of this study broadens the capabilities in terms of tool, process and expertise access for potential industrial partners. Thus, an important milestone has been reached in aligning the three research institutes to offer advanced AI systems with novel non-volatile memory components.

Acknowledgements

This study was fully financed by TEMPO project. The TEMPO project has received funding from the Electronic Components and Systems for European Leadership Joint Undertaking under grant agreement No 826655. This Joint Undertaking receives support from the European Union's Horizon 2020 research and innovation program and Belgium, France, Germany, Switzerland, The Netherlands.

References

[1] C. Bigot, A. Danel, S. Thevenin (2005). Influence of Metal Contamination in the Measurement of p-Type Cz Silicon Wafer Lifetime and Impact on the Oxide Growth. Solid State Phenomena (Vols. 108-109), S. 297–302 doi:10.4028/www.scientific.net/SSP.108-109.297

[2] Y. Borde, A. Danel, A. Roche, A. Grouillet, M. Veillerot (2007). Estimation of Detrimental Impact of New Metal Candidates in Advanced Microelectronics. Solid State Phenomena (Vol. 134), S. 247–250 doi:10.4028/www.scientific.net/SSP.134.247

[3] F. Boulard, V. Gros, C. Porzier, L. Brunet, V. Lapras, F. Fournel, N. Posseme (21. Mai 2022). Bevel contamination management in 3D integration by localized SiO2 deposition. SSRN Journal (SSRN Electronic Journal)

[4] D. Autillo, et al. (June 2020). FR Patentnr. U.S. Patent No 20200203190 A1

9

A Framework for Integrating Automated Diagnosis into Simulation

David Kaufmann and Franz Wotawa

Graz University of Technology, Austria

Abstract

Automatically detecting and locating faults in systems is of particular interest for mitigating undesired effects during operation. Many diagnosis approaches have been proposed including model-based diagnosis, which allows to derive diagnoses from system models directly. In this paper, we present a framework bringing together simulation models with diagnosis allowing for evaluating and testing diagnosis models close to its real world application. The framework makes use of functional mock-up units for bringing together simulation models and enables their integration with ordinary programs written in either Python or Java. We present the integration of simulation and diagnosis using a two-lamp example model.

Keywords: model-based diagnosis, fault detection, fault localization, physical simulation.

9.1 Introduction

To keep systems operational, we need to carry out diagnoses regularly. Diagnosis includes the detection of failures, the localization of corresponding root causes, and repair. We carry out regular maintenance activities that include diagnosis and predictions regarding the remaining lifetime of components to prevent systems from breaking during use. However, there is no guarantee

DOI: 10.1201/9781003377382-9

that system components are not breaking during operation, even when carrying out maintenance as requested. In some cases, it is sufficient to indicate such a failure, i.e., via presenting a warning or error message and passing mitigation measures to someone else. Unfortunately, there are systems like autonomous systems where we can hardly achieve such a mitigation process. For example, in fully autonomous driving, there is no driver anymore for passing control. Therefore, there is a need for coming up with advanced diagnosis solutions that cover detection, localization, and repair. A practical real world problem demonstration of an on-board control agent was validated in the year 1999, within the scope of Deep Space One, a space exploration mission, carried out by NASA. Regarding this, the authors of the paper [4] describe developed methods related to model-based programming principles, including the area of model-based diagnosis. The methods were applied on autonomous systems, designed for high reliability, operating as subject of a spacecraft system.

When we want to integrate advanced diagnosis into systems, we need to come up with means for allowing us to easily couple monitoring with diagnosis. As stated by the authors in [3], the coupling enables the diagnosis method to detect and localize faults based on observations, obtained by monitoring a cyber-physical system (CPS). Furthermore, we require close integration of today's development processes, which rely on system simulation. The latter aspect is of uttermost importance for showing early that diagnosis based on monitoring can improve the overall behaviour of a system even when working not as expected. We contribute to this challenge and present a framework for integrating different simulation models and diagnoses. The framework utilizes combining functional mock-up units (FMUs) that may originate from modeling environments like OpenModelica[1] with ordinary programming languages like Java or Python. We use these language capabilities to integrate diagnosis functionality. The architecture of our framework is based on the client-server pattern and implemented using Docker containers.

Using our framework, we can easily add diagnoses into systems. In addition, we can use this framework for carrying out verification and validation of the system functionality enhanced with diagnosis capabilities. In this manuscript, we present the framework and show the integration of diagnosis. For the latter purpose, we make use of a simple example. We will make the framework and the underlying diagnosis engine available for free and as open-source. The framework contributes to research area of Edge Artificial

---

[1]see https://openmodelica.org

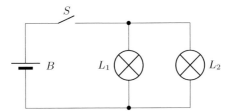

Figure 9.1 A simple electric circuit comprising bulbs, a switch and a battery.

Intelligence because it enables the direct use of diagnosis functionality that is based on Artificial Intelligence methodology in systems without the necessity for communication with other systems.

We structure the paper as follows. First, we discuss the foundations behind the used diagnosis method, i.e., model-based diagnosis. Afterwards, we describe the simulation framework that is based functional mock-up units using a small example. We further show how diagnosis can be integrated into this framework, and finally we conclude the paper.

9.2 Model-based Diagnosis

Diagnosis, i.e., the detection of failures and the identification of faults, have been of interest for several decades. In the early eighties of the last century, Davis and colleagues [1][2] introduced the basic concepts behind model-based diagnosis. The idea is to utilize a model of the system directly for detecting and locating faults. Reiter [5] formalized the idea utilizing first-order logic. For a more recent paper we refer to Wotawa and Kaufmann [8] where the authors introduced how advanced reasoning systems can be used for computing diagnosis. For recent applications of diagnosis in the context of CPS have a look at [3][9][7][6].

In the following, we illustrate the basic concepts using a small example circuit comprising a battery B, a switch S, and two bulbs L_1, L_2. The bulbs are put in parallel and both should provide light when the switch is turned on and the battery is not empty. Otherwise, both bulbs do not deliver any light. We depict the circuit in Figure 9.1. If we know that the switch S is on, and the battery is working as expected, then we also would expect both bulbs to be illuminated. In case one bulb is emitting light but the other is not, we would immediately to derive that the bulb with no transmitting light is broken.

To compute diagnoses from system models, we first need to come up with a model of the system that we want to diagnose. Such models comprise components and their connections, via ports. Hence, in the following, we discuss the component models, and a model of connections separately. For the electric circuit, we simplify modelling by only considering that components like batteries are providing electrical power, some are transferring power like switches, and others are consuming power. Furthermore, we utilize first-order logic for formalization where we follow Prolog syntax[2]. For all the component models we describe how values are computed assuming that the component is of a particular type and that it is working as expected. For the type we use a predicate type\2 and for stating the component to be correct a predicate nab\1.

Battery A component X that is a battery is when working correctly providing a nominal power at its output.

val(pow(X),nominal) :- type(X,bat), nab(X).

Switch A component X that is a switch works as follows. If it is on and working as expected, then the output must have the same value as the input port and vice versa. If it is off, the switch is not transferring any power.

val(out_pow(X),V) :-   type(X,sw), on(X),
                  val(in_pow(X),V), nab(X).
val(in_pow(X),V) :-   type(X,sw), on(X),
                  val(out_pow(X),V), nab(X).
val(out_pow(X),zero) :- type(X,sw), off(X), nab(X).

Lamp A lamp X is on, whenever there is a power on its input. If it emits light, then there must be power on its input. If there is no power at the input of X, then the light must be off.

val(light(X),on) :- type(X,lamp), val(in_pow(X),
               nominal), nab(X).
val(in_pow(X), nominal) :- type(X,lamp),
                  val(light(X),on).
val(light(X),off) :-   type(X, lamp),
                  val(in_pow(X),zero), nab(X).

---

[2]We are using Prolog syntax because recent solvers like Clingo (see https://potassco.org/clingo/) are relying on it.

For completing the model, we introduce connections using a predicate conn\2 that allows to state two ports to be connected. The behaviour of a component comprises the transfer of values in both directions, and stating that it is impossible to have different values at a connection. The following rules are covering this behaviour:

val(X,V) :- conn(X,Y), val(Y,V).
val(Y,V) :- conn(X,Y), val(X,V).

:- val(X,V), val(X,W), not V=W.

To use a model for diagnosis we only need to define the structure of the system making use of the component models. For the two bulb example, we define a battery, a switch, and two bulbs that are connected accordingly to Figure 9.1.

type(b, bat).
type(s, sw).
type(l1, lamp).
type(l2, lamp).

conn(in_pow(s), pow(b)).
conn(out_pow(s), in_pow(l1)).
conn(out_pow(s), in_pow(l2)).

To use this model for diagnosis, we further need observations. We might state that the switch s is on, bulb l1 is not on but l2 is. Again we can make use of Prolog to represent this knowledge as facts:

on(s).
val(light(l1),off).
val(light(l2),on).

When using a diagnosis engine like described in [8] we obtain one single fault diagnosis {l1}. But how is this working? The diagnosis engine makes use of a simple mechanism. It searches for a truth setting to the nab\1 predicates, such that the model together with these assumptions is not leading to a contradiction. When assuming l1 to be not working, the fact that lamp l2 is on can be derived. However, we cannot derive anything else that would lead to a contradiction.

Note that this simple model is also working in other more interesting cases. Let us assume that the switch is on but no light is on. For this case, the diagnosis engine delivers three diagnoses: $\{b\}$, $\{s\}$, and $\{l1, l2\}$ stating the either the battery is empty, the switch is broken, or both lamps are not working at the same time. Another interesting case that might occur is setting the switch to off, put still one lamp, i.e., $l1$ is on. In this case we only obtain a double fault diagnosis $\{s, l2\}$ stating that the switch is not working as expected and lamp $l2$ as well.

9.3 Simulation and Diagnosis Framework

In the following section, we introduce a framework making use of two collaborating tools, comprising a simulation environment for function mock-up unit (FMU)[3] models and a diagnose application based on the theorem solver Clingo[4]. Figure 9.2 gives a brief overview of the framework and the operating principles. The FMU simulation tool server is utilized to run a CPS model within the given simulation environment, whereas the client enables to control the simulation. The separation enables to execute other applications, tools and methods after each simulation time step update, as the ASP Diagnose Tool (see Section 9.3.2). The mentioned tool receives the observations provided by the simulation framework and a settings configuration to compute the diagnose of a system, based on an abstract model, developed with the declarative programming language ASP (Answer Set Programming). Further, the diagnose may be used to control the inputs and parameter to restore a safe operating system or to bring the system in a state to prevent harm to the system or environment.

9.3.1 FMU Simulation Tool

The developed application provides an entire environment to load, configure, run, observe and control simulations related to CPS models. In general, the application is set up as a client-server system to distribute the structure between the provider of a service, the server, and the service requester, the client. The service executed on the server is defined as the simulator environment providing the options to observe and control the simulation by client requests during run-time. The reason of using a client-server

---

[3] see https://fmi-standard.org
[4] see https://potassco.org/clingo/

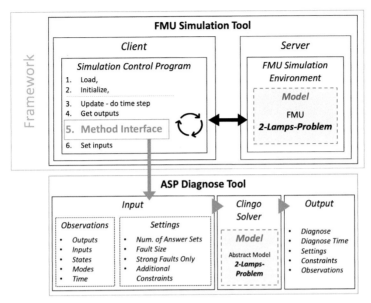

Figure 9.2 Illustration of the simulation and diagnose environment as well as the overall operating principles. The framework of the FMU Simulation Tool provides an interface to enable the integration of a diagnose tool and/or other methods. The models can be substituted by any others in the provided framework.

system is to detach the simulation environment and the observation/control process. The separation enables the user to utilize individual programming environments/languages as client, whereas the server works independent to the selected client environment, receiving and sending the control commands and simulation observations via a REST (Representational State Transfer) application programming interface. In order to run a simulation of a CPS model with the described application, a fundamental requirement is to generate a standardized FMU from the given model. Common modeling software as OpenModelica or Matlab[5] have a FMU generation tool already implemented, but there are also other applications, as e.g. UniFMU [6], which are capable of generating a FMU from different language source code (Python, Java or C/C++). A FMU enables to use a general simulation environment for all kind of models, although they are build on different sources. The simulation environment is developed to execute a step by step

---

[5]see https://de.mathworks.com/products/matlab.html
[6]see https://github.com/INTO-CPS-Association/unifmu

(for a given time step) simulation. To enable that feature, it is essential that the FMU is generated as a co-simulation model. Within a co-simulation setup, the numerical solver is embedded and supplied by the generated FMU. By the provided interface methods, the FMU can be controlled by setting the inputs and parameter, computing the next simulation time step, and reading the resulting observations. The given setup, enables to execute tools and methods while the simulation is paused after a simulated time step.

9.3.2 ASP Diagnose Tool

To enable diagnoses based on observations of a given CPS model, we developed a diagnose tool. This tool is built up on the theorem solver Clingo and makes use of the provided methods within a Python environment. In addition the tool provides extended functionalities, e.g. including observations as simulation outputs, inputs, states, modes or time and applying optional settings as limiting the number of required answer sets, setting the maximum fault size search space for abnormal component behaviour, considering additional fault modes and adding other constraints to be considered.

The tool is designed to iterate through each fault size in ascending order, whereas fault size zero indicates a normal operating system without detecting any abnormal behaviour in the diagnosed components. The procedure is repeated for each fault size, except when the model is satisfied for fault size zero, what is interpreted as no abnormal component is present for the given observation. The detailed theorem solver implementation structure is shown in algorithm 1, which was initially introduced and applied by Wotawa, Nica and Kaufmann [3]. In the following, we briefly describe the setup of the stated algorithm. First the input model is initiated, defined as an abstract model (M), comprising the system description (SD), observations (Obs) and additional fault modes (FM) to guide the diagnosis search. We start with an empty diagnosis set (DS) and compute diagnosis of a certain size, iterating from 0 to n. Line 4 shows how the limitation of the number for abnormal predicates is applied to the model (M_f), before the solver is called (line 5). A specified answer set is returned and filtered for abnormal predicates (S) only. To prevent the multiple occurrence of abnormal elements (C) in the iterations, the corresponding integrity constraints are added to the model (M_f) as stated in line 12. In relation to the given example in Section 9.2, a integrity constraint at fault size 1 could be stated as `:- ab(l1).` for a detected abnormal behaviour of the component lamp 1.

Besides the main diagnose algorithm, the tool enables different output options to simplify the evaluation of the received diagnose. Thus, the received data can be exported in a JSON file, CSV file or directly printed in the terminal during run-time. The output results are the detailed computed diagnose, the total number of found diagnosis for each fault size, an indicator for strong faults and the diagnose time separated for each fault size and in total. As input, the tool requires the Prolog model, representing the CPS as abstract model (see Section 9.2), and the related observation/constraint file with all necessary input information to execute the diagnose process.

In reference to Figure 9.2, we show the simulation tool update loop, where an update is triggered and the observations are received. Further the observations are passed by the method interface as input to the implemented diagnose tool. Before calling the diagnose, some configurations are specified, as the abstract model, the maximum number of computing answer sets, the maximum fault size of interest and the observations, which are generated based on the simulation output information. In addition, the diagnose output format, e.g., JSON or CSV can be selected. Last, the ASP theorem solver with the given model, configuration and simulation observations is executed. After receiving the diagnose result of the current time frame, it is stored in the defined format structure and the simulation is continued with the next time step in the loop.

9.4 Experiment

To show the applicability of the framework, we make use of the two-lamps-model concept as shown in Figure 9.1. For the simulation, a model of the two-lamps-model (see Listing 1) is generated in OpenModelica comprising a battery ($5.0V$), a closing switch and two light bulbs (100Ω). Besides the connection of each component, the model also describes inputs, which can be set during the simulation. These inputs are covering the fault type of each component and the operational switch logic. To give an example of the component programming, the switch model is shown in more detail at Listing 2. Besides the component mode, the equations also represent the behaviour based on different fault states, e.g. a broken switch, resulting in an infinite high internal resistor value equal to an open electrical circuit. An equivalent fault state is implemented for each component as shown in Table 9.1.

Moreover the OpenModelica model is converted into a co-simulation FMU, which enables to use the model in the described FMU simulation tool.

Algorithm 1 ASPDiag(SD, Obs, FM, n)

For a more detailed description of the algorithm see [3].

Input: An ASP diagnosis model M, and the desired cardinality n
Output: All minimal diagnoses up to n

1: Let DS be $\{\}$
2: Let M_f be M.
3: **for** $i = 0$ **to** n **do**
4:     $M'_f = M_f \cup \{ :\text{- not numABs}(i). \}$
5:     $S = \mathcal{F}\left(\textbf{ASPSolver}(M'_f)\right)$
6:     **if** i is 0 **and** S is $\{\{\}\}$ **then**
7:        **return** S
8:     **end if**
9:     Let DS be $DS \cup S$.
10:    **for** Δ in S **do**
11:       Let $C = AB(\Delta)$ be the set $\{c_1, \ldots, c_i\}$
12:       $M_f = M_f \cup \{ :\text{- ab}(c_1), \ldots, \text{ab}(c_i). \}$.
13:    **end for**
14: **end for**
15: **return** DS

```
model Two_Lamp_Circuit
  PhysicalFaultModeling.PFM_Bulb bulb1(r = 100.0);
  PhysicalFaultModeling.PFM_Bulb bulb2(r = 100.0);
  PhysicalFaultModeling.PFM_Switch sw;
  PhysicalFaultModeling.PFM_Ground gnd;
  PhysicalFaultModeling.PFM_Battery bat(vn = 5.0);
equation
  connect(gnd.p, bat.m);
  connect(bat.p, sw.p);
  connect(sw.m, bulb1.p);
  connect(sw.m, bulb2.p);
  connect(bulb1.m, gnd.p);
  connect(bulb2.m, gnd.p);
end Two_Lamp_Circuit;

model Two_Lamp_Circuit_Testbench
  PhysicalFaultModeling.Two_Lamp_Circuit sut;
  input FaultType bat_state(start=FaultType.ok);
  input OperationalMode switch_mode(start=OperationalMode.close);
  input FaultType switch_state(start=FaultType.ok);
  input FaultType bulb1_state(start=FaultType.ok);
  input FaultType bulb2_state(start=FaultType.ok);
equation
  sut.sw.mode = switch_mode;
  sut.bat.state = bat_state;
  sut.sw.state = switch_state;
  sut.bulb1.state = bulb1_state;
  sut.bulb2.state = bulb2_state;
end Two_Lamp_Circuit_Testbench;
```

Listing 1 OpenModelica model of a two-lamp electrical circuit with fault injection capability to each used component. The component connections are specified to describe the same electrical circuit as given in Figure 9.1.

```
model PFM_Switch
  extends PhysicalFaultModeling.PFM_Component;
  PhysicalFaultModeling.OperationalMode mode(start=OperationalMode.open);
  Modelica.Units.SI.Resistance r_int;
equation
  v = r_int * i;
  if state == FaultType.ok then
    if mode == OperationalMode.open then
      r_int = 1e9;
    else
      r_int = 1e-9;
    end if;
  elseif state == FaultType.broken then
    r_int = 1e9;
  else
    r_int = 1e-9;
  end if;
end PFM_Switch;
```

Listing 2 OpenModelica model of a switch component including a mode {open, close} and fault state {ok, broken, short} implementation logic.

Table 9.1 CPS Model component state description for the light bulb, switch and battery. All used states, including fault states of the components are shown.

Component	State	Description
light bulb (bulb), switch (sw)	ok	ordinary behaviour
	broken	open connection in eletrical circuit
	short	short in the electrical circuit
battery (bat)	ok	ordinary behaviour
	empty	empty battery fault

In order to simulate the model behaviour in detail, the update time step is set to 0.01 seconds. In addition, the fault injection during run-time is configured to trigger a single light bulb fault at 0.2 seconds and a switch fault after 0.3 seconds, which is described in detail at the simulation part of Figure 9.4.

For the diagnose part, we make use of the described abstract model of the electrical two-lamps circuit (see Section 9.2). The overall framework is built up in a way, that a diagnose is computed after each simulated time step and is based on the actual observations (simulation outputs, parameter and inputs). The use of a co-simulation FMU, allows a step-by-step simulation, which enables to pause the simulation during the diagnose process and continuing afterwards. Therefore, the diagnose time effort has no impact on the overall simulation results.

Figure 9.3 shows the observed signals for the current flow in the battery, light bulb 1 and 2 as well as the actual switch mode. Further the injected faults are highlighted at the correlated time point. In Figure 9.4 a table represents the observations for the three interesting time sections, as the normal behaviour, a broken light bulb and a broken switch. After reaching simulation time 0.05

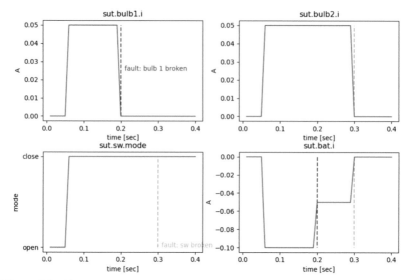

Figure 9.3   Simulation showing the measured signal output of the two bulbs, switch and the battery. For this example a fault injection (`broken`) in bulb 1 after 0.2 seconds (red indicator) and a fault injection (`broken`) to the switch after 0.3 seconds (orange indicator) is initiated.

seconds, the switch mode is changed from `open` to `closed` and the model shows the expected ordinary behaviour without any abnormal components. Both light bulbs are operating at an expected current consumption of 0.05 A. These observations are translated to a readable input format for the diagnose tool, which is shown in the corresponding status row "Observation" (see Figure 9.4). In regards to the abstract model and the observation input, the diagnose tool computed a satisfied model at fault size zero, which concludes an expected ordinary behaviour of all considered components.

The time section at 0.2 seconds shows the behaviour with a broken light bulb. Thus the current consumption of bulb 1 immediately drops to 0.0 A and the diagnose observation changes from mode `on` to `off`. Since the main power switch is still closed and bulb 2 is in active mode `on`, the diagnose model concludes component bulb 1 as abnormal `ab(l1)`. The next investigated fault (`broken`) is injected to the closed switch. Since the power supply for both light bulbs is not given, the current consumption drops to 0.0 A. The diagnose model concludes as expected an abnormal switch (`ab(s)`) or battery (`ab(b)`) based on the given observations for single faults. Under consideration of double faults, the computed diagnose

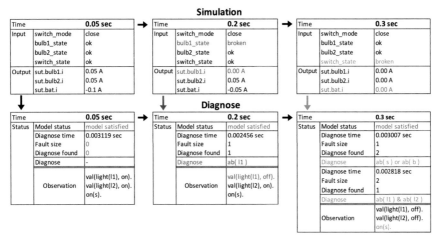

Figure 9.4 Simulation and diagnose output results based on the electrical two-lamps circuit with a broken bulb after 0.2 seconds and a broken switch at 0.3 seconds. The upper tables illustrate the simulation input/output signals, which are used as observation for the diagnose (lower tables) part. Based on the given observations for the three selected time steps, different diagnose results are obtained.

shows a combination of an abnormal behaviour of light bulb 1 and bulb 2 (`{ab(l1), ab(l2)}`), which is also a possible solution for the given observation.

9.5 Conclusion

In this paper, we have shown how to use an automated diagnosis method within a simulation framework for a CPS (cyber-physical system). For this purpose we introduced the foundations behind the model-based diagnosis method based on a simple electric circuit model comprising two light bulbs, a switch and battery. Next we describe a framework for simulating the developed CPS model with the ability of fault injection during run-time. In order to run the model in the given framework, it is essential to generate a functional mock-up unit (FMU) based on the developed electrical two lamp circuit model. By providing the FMU in co-simulation configuration, the simulation can run in a step-by-step mode (time steps), which enables to call other functions, as for example the diagnose method, while the simulation is paused and continued with the next time step.

Besides the physical electrical circuit model, an abstract model for diagnosis is developed in the declarative programming language Prolog. For computing the diagnose based on observations of the model simulation, we introduce a tool which uses the theorem solver Clingo and offers additional productive options. The tool is developed to automate the process of searching for abnormal components at each fault size (in ascending order). To prevent multiple occurrence of abnormal components in higher fault sizes, the derived results are continuously added as constraints to the model.

In order to demonstrate the concept of the simulation framework with the automated diagnose tool, we executed an experiment based on the described electrical two-lamps circuit model with the capability of fault injection to the light bulbs and switch. After each time step of simulation, the received observations are forwarded as input to the diagnose tool. The diagnose tool enables to detect the injected faults in a fast and accurate way, as a single bulb fault or even the more interesting case, when a switch erroneously indicates a closed position although both light bulbs are powered off. In this case, we obtain a single fault for an abnormal switch or battery behaviour, and a double fault stating an abnormal behaviour for both light bulbs in combination.

For the purpose of deploying the diagnose tool on a system applied under real environmental conditions, validation and verification is a fundamental process. Thus we make use of a simulated environment framework, enabling a high test case coverage of scenarios with abnormal component behaviour of the system under test. In addition, the required time to conclude a diagnose, may also lead to issues and need to be considered in the evaluation. Future research includes investigating more complex CPSs by making use of the discussed simulation framework in combination with the diagnose tool and further development of both tools.

Acknowledgments

The research was supported by ECSEL JU under the project H2020 826060 AI4DI - Artificial Intelligence for Digitising Industry. AI4DI is funded by the Austrian Federal Ministry of Transport, Innovation and Technology (BMVIT) under the program "ICT of the Future" between May 2019 and April 2022. More information can be retrieved from https://iktderzukunft.at/en/ bm⦿ⓘ.

References

[1] R. Davis, H. Shrobe, W. Hamscher, K. Wieckert, M. Shirley, and S. Polit. Diagnosis based on structure and function. In *Proceedings AAAI*, pages 137–142, Pittsburgh, August 1982. AAAI Press.

[2] R. Davis. Diagnostic reasoning based on structure and behavior. *Artificial Intelligence*, 24:347–410, 1984.

[3] D. Kaufmann, I. Nica, and F. Wotawa. Intelligent agents diagnostics - enhancing cyber-physical systems with self-diagnostic capabilities. *Adv. Intell. Syst.*, 3(5):2000218, 2021.

[4] N. Muscettola, P. Pandurang Nayak, B. Pell, and B. C. Williams. Remote agent: to boldly go where no ai system has gone before. *Artificial Intelligence*, 103(1):5–47, 1998. Artificial Intelligence 40 years later.

[5] R. Reiter. A theory of diagnosis from first principles. *Artificial Intelligence*, 32(1):57–95, 1987.

[6] F. Wotawa. Reasoning from first principles for self-adaptive and autonomous systems. In E. Lughofer and M. Sayed-Mouchaweh, editors, *Predictive Maintenance in Dynamic Systems – Advanced Methods, Decision Support Tools and Real-World Applications*. Springer, 2019.

[7] F. Wotawa. Using model-based reasoning for self-adaptive control of smart battery systems. In Moamar Sayed-Mouchaweh, editor, *Artificial Intelligence Techniques for a Scalable Energy Transition – Advanced Methods, Digital Technologies, Decision Support Tools, and Applications*. Springer, 2020.

[8] F. Wotawa and D. Kaufmann. Model-based reasoning using answer set programming. *Applied Intelligence*, 2022.

[9] F. Wotawa, O. A. Tazl, and D. Kaufmann. Automated diagnosis of cyber-physical systems. In *IEA/AIE (2)*, volume 12799 of *Lecture Notes in Computer Science*, pages 441–452. Springer, 2021.

10

Deploying a Convolutional Neural Network on Edge MCU and Neuromorphic Hardware Platforms

Simon Narduzzi[1], **Dorvan Favre**[1,2], **Nuria Pazos Escudero**[2]
and L. Andrea Dunbar[1]

[1]CSEM, Switzerland
[2]HE-Arc, Switzerland

Abstract

The rapid development of embedded technologies in recent decades has led to the advent of dedicated inference platforms for deep learning. However, unlike development libraries for the algorithms, hardware deployment is highly fragmented in both technology, tools, and usability. Moreover, emerging paradigms such as spiking neural networks do not use the same prediction process, making the comparison between platforms difficult. In this paper, we deploy a convolutional neural network model on different platforms comprising microcontrollers with and without deep learning accelerators and an event-based accelerator and compare their performance. We also report the perceived effort of deployment for each platform.

Keywords: neuromorphic computing, IoT, kendryte, DynapCNN, STM32, performance, comparison, benchmark.

10.1 Introduction

Edge computing is a key tool in harnessing the possibilities of artificial intelligence. Some advantages of edge over cloud processing are low latency, allowing real-time application and connectivity independence, i.e., no need

DOI: 10.1201/9781003377382-10

of infrastructure and no transmission of sensitive data, allowing improved security and privacy-preserving applications. However, perhaps the most important and as yet untapped potential of edge computing is in the low power possibilities. Low power allows always-on IoT devices for seamlessly integrated intelligent systems. Creating edge-based IoT devices often requires limited hardware resources, both in terms of power and on-device memory. Today's intelligence is mainly based on Deep Learning (DL) networks which are power and memory hungry. This conflict has resulted in several emerging technologies and platforms to perform efficient inference at the edge.

Established companies have both targeted the IoT device by creating ultra-low-power processors (Intel Loihi, STM32 Cortex-M4), but there are also several other innovative platforms such as DynapCNN[1] and Kendryte K210[2] specialized for deep neural network inference with a very little power budget. The specialized nature and variety of products and platforms require platform-specific software tools, making the deployment of one model on several platforms cumbersome and creating a barrier to technology adoption. Moreover, the lack of hardware standardization coupled with the necessary customization of the software makes it difficult to compare, and thus choose, the best technology.

To remove this barrier, it is essential to facilitate access to platforms to non-hardware experts. Indeed, the success of DL is essentially linked to the acceleration provided by graphical processing units (GPUs). Currently, only a very small proportion of users have mastered the CUDA programming language used by the majority of GPUs. In most DL libraries, mobilization of the necessary resources can be called in a single command line, without the user having to understand the technology behind it. This kind of single instruction would empower the data scientists in the porting to edge devices.

In this short paper, we give a brief summary of works that address the challenges of implementing DL on different hardware platforms. Initially, we present our results on a basic neural network deployment on edge devices, and then we compare the performance of 3 selected devices. Finally, we describe the lessons learned and present solutions to facilitate the deployment of these models in the future.

10.2 Related Work

Benchmarking low-resource platforms is a necessary process to select the best platforms to embed algorithms. It is a tricky procedure, as the performance of a platform depends on several aspects: the available memory and processing units, the technology of the hardware, and the frameworks and tools used during the deployment of the models to benchmark. To harmonize the performance assessment, benchmarking suites such as TinyMLPerf [3] have been created. Recently, a benchmarking suite has been developed for event-based neuromorphic hardware [4]. However, both these solutions still need manual adaptation of the code to run on new platforms. While the benchmarking gives good insights about which and why to select a certain platform. It still remains the question of how to use the benchmarking tools itself. Each platform comes with its own SDK, conversion tools, and constraint of utilization, which in turn limits the possibility of comparing the platforms between them.

Today, many benchmarks are therefore performed on just a few hardware platforms and comparing only a single use-case, as alternatives are more cumbersome. Furthermore, it is easier to benchmark and compare platforms from the same constructor, as the deployment pipelines are usually similar between devices. In this regard, standard architectures LeNet-5 and ResNet-20 have been benchmarked on a few STM32 boards [5]. Machine learning algorithms have also been compared on Cortex-M processors [6][7]. Some efforts of cross-constructor benchmarking have also been made. For example, a recent work deployed a gesture recognition and wake-up words application on an Arduino Nano BLE and a STM32 NUCLEO-F401RE [8] using a convolutional neural network.

While the above research focuses on the established STM32 Cortex-M based MCUs, some emerging processors are also explored [9], but the research in this domain remains scarce. Furthermore, the deployment pipelines are not documented, which limits the reproducibility of the results. In our research, we deploy a single neural network on three different platforms and observe their performance. We also highlight the difference between the deployment pipelines of each constructor, and we perform a qualitative study of the easiness of deployment on each system.

10.3 Methods

In this section, we present the selected task and associated experimental setup, and a method to evaluate the effort of the deployment.

10.3.1 Neural Network Deployment

In our experiment, we use 3 different boards. We select boards from different constructors to show the (large) variety of tools and processing available in

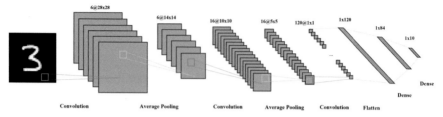

Figure 10.1   Illustration of LeNet-5 architecture.

edge devices today. These sample devices are a very small subset of the large variety of devices today, but they show that with only three different board manufacturers, an extensive adaptation of the deployment pipeline is necessary. The selected 3 devices for our experiments are the following: a Kendryte K210 from Canaan, a dual-core RISC-V processor with floating-point units; an STM32L4R9 from STMicroelectronics (ST) with an ARM Cortex-M4 core also including floating-point unit, and SynSense DynapCNN, an event-based processor. Table 10.1 summarizes the major differences between these platforms.

10.3.1.1 Task and Model

We tested the selected platforms on a simple LeNet-5 [10] networks trained on MNIST, which architecture is displayed in Figure 10.1. This architecture, composed of convolutions layers, average pooling and dense layers, is compatible with all selected platforms. The architecture was trained for 30 epochs with a learning rate $1e - 4$. Tensorflow 2.9.1 was used to define the H5 model running on the Sipeed and ST boards, while PyTorch 1.11.0 was used for DynapCNN. Unfortunately, our efforts to transfer the weights from the Tensorflow model to the PyTorch failed, and we had to train the models separately. The Keras and PyTorch models reached an accuracy of 99.44% and 99.38% on the train set, respectively. We perform inference on the first 1000 images of the test dataset.

10.3.1.2 Experimental Setup

For each platform, we used the latest tools available at the time at which this article was written.

Kendryte K210

The Kendryte K210 is used with the Sipeed MaixDock M1. The Neural networks embedded in this device were converted from Keras H5 file format, using Tensorflow 2.9.1 and associated TFLite. The firmware version of the Kendryte is 0.6.2, and the version of the NNCase package used for conversion is 0.2.

STM32L4R9

The STM32L4R9 board with an Arm Cortex-M4 core processor from ST is programmed in C. Due to the complexity of hardware initialization, ST provides a tool, STM32CubeMX 6.5.0, which automatically generates an initial C project for a specific board. The tool X-CUBE-AI 7.1.0 converts TFLite models into C files which are, alongside the X-CUBE-AI inference library, added to the project. The Keras H5 file network is converted to TFLite format using Tensorflow 2.8.2 and Python 3.6. Gcc-arm-none-eabi 15:10.3-2021.07-4 and Make 4.2.1 are used to compile the whole project, and STM32CubeProgrammer 2.10.0 is used to upload the binaries on the device.

DynapCNN

The SynSense DynapCNN processor was programmed using Python 3.7.13 with PyTorch 1.11.0, Sinabs 0.3.3 (and underlying Sinabs-DynapCNN 0.3.1.dev3), and Samna 0.14.33.0 libraries. The neural network is written in PyTorch and converted to a spiking version using Sinabs, while Samna is used to map the network to the hardware. The inputs are presented to the network using a preprocessing function that generates spikes[1] from random sampling of the image, using the following function, where tWindow is the duration of the spiking frame and img has shape [channels, width, height]:

```
def to_spikes(img, tWindow=100):
    rnd = (np.random.rand(self.tWindow, *img.shape)
    img = rnd < img.numpy()/255.0).astype(float)
    return torch.from_numpy(img).float()
```

During our simulation, we found 100 timesteps to be sufficient to reach equivalent accuracy between the spiking and non-spiking version of MNIST.

10.3.1.3 Deployment

For standalone platforms, the network was converted and uploaded to the platform. For Kendryte, the inference script was written such that the model

---

[1]Spikes are binary events (on or off) distributed in input space and time.

Table 10.1   Relevant technical specifications of the devices (from constructor websites).

Board	Kendryte K210	STM32L4R9	DynapCNN
Processor ISA	Dual-core RISC-V 64b	ARM Cortex-M4	Event-based
Power Consumption	300mW	66mW	1mW
Max Frequency (MHz)	900	120	-
TOPS/W	3.3	-	-
Standalone	Yes	Yes	No
Event-based	No	No	Yes
Language	MicroPython	C	Python

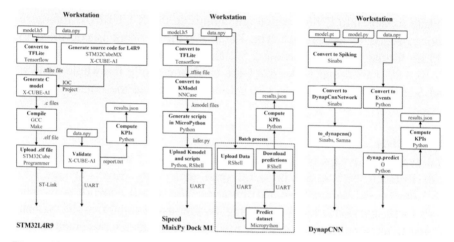

Figure 10.2   Deployment pipelines for all platforms. From left to right: STM32L4R9, Kendryte K210 and DynapCNN. For DynapCNN, the pipeline is contained in a single Python script, while the other relay on external languages and tools.

is loaded at the beginning of the script and processes images one by one. The images are transmitted via serial communication and inferred by inference script. In X-CUBE-AI, this is automatically done, while Kendryte requires a script that sends batches of images and obtains the predictions. For DynapCNN, the images are predicted by sending the corresponding events to the device and reading the output events from the buffer of the board.

The prediction time is provided automatically by the X-CUBE-AI platform, while Kendryte requires to time the prediction manually. In the MicroPython script used for inference on Kendryte, we put a counter around the line performing the inference. For DynapCNN, the reported times corresponds to the timestamp of the first output event and the final output event, respectively. Both times are averaged over the test samples. The computation of the key performance indicators (accuracy, mean time) is performed offline. Figure 10.2 illustrates the pipelines for all platforms.

10.3.2 Measuring the Ease of Deployment

One of the major criteria for the adoption of a product is the ease of use, meaning how much one user is autonomous in using the device. This highly depends on the user skills, but also on the quality of the documentation. For embedded machine learning, the documentation should explicitly describe the procedure to deploy a model once the user receives the new platform. We have identified 5 different phases that are required when using a microcontroller product for AI acceleration.

- **Acquisition (A)**: this phase comprises the effort needed to place an order for the device and the time necessary to ship the device. A small effort would correspond to ordering the platform from a website and receiving it within the next week. A large effort requires to contact the company by phone or email and wait for two month to receive the device.
- **Setup (S)**: this phase comprises the effort needed to install the required environment. A small effort would require installing a python package from pip or an executable available from the constructor website. A large effort requires installing multiple packages which versions depend on the firmware of the device or the version of Python packages used to train the model, as well as dependencies on external tools.
- **Getting started (G)**: this phase is the effort needed to replicate the examples given in the documentation. A small effort would correspond to a full deployment example done within one hour. A large effort would require support from the constructor.
- **Model preparation (M)**: this phase comprises the effort needed to convert a PyTorch/Tensorflow model to the proprietary format of the device. A small effort would correspond to a single command line with arguments. A large effort corresponds to manually writing the neural network in the proprietary format and transferring the weights, with limited help from the conversion tool, or requiring intervention from the constructor.
- **Inference (I)**: this phase comprises the effort needed to perform inference once the model is embedded to the device. A small effort would correspond to a single command line or instruction to perform inference, a medium effort requires writing an inference script and deploying it manually on the hardware platform. A large effort would require intervention from the constructor.

Each phase is assigned with a number between 1 and 5. The total score represents the complexity of deployment. A low value (5) corresponds to a small effort necessary to deploy a model on a never-used platform, while 25 corresponds to a large effort.

10.4 Results

In this section, we present the results and metrics recorded for each platform, and the effort perceived by the team to perform the experiments.

10.4.1 Inference Results

The models were successfully deployed on all platforms. Table 10.2 summarizes the results on the 1000 first samples of MNIST test dataset. It can be observed that the balanced accuracy is not homogeneous between the platforms. This difference is certainly caused by the different transformations affecting the models during the deployment (conversion). While we initially tried to deploy full-precision models and a quantized version of them, we only had time to deploy it on the ST platform. The evaluation of quantized-aware trained models and evaluation DynapCNN and Kendryte K210 using integer weights is a future work. The models run faster when using 8-bit integer precision on STM32 (even if the platform is made to compute 32-bit floats). The Kendryte K210 is the fastest to compute synchronous frames while DynapCNN is the fastest to provide a result in a 32-bit precision, with 98.79% precision using only the first spike[2]. Unfortunately, only the DynapCNN provides an estimation of the energy consumption, obtained with Sinabs by computing the average number of synpatic operations over the course of the simulations. All the metrics are averaged over the test partition.

Table 10.2   Results on MNIST dataset for all platforms. For the DynapCNN, we report the accuracy and latency for the first spike prediction and over the entire simulation.

Platform	Kendryte K210	STM32L4R9		DynapCNN
Bit Precision	float-32	float-32	int-8	float-32
Size (KB)	94.2	359.2	90.5	-
Accuracy	97.23%	98.26%	94.07%	98.79% / 99.09%
Latency (ms)	54.17	80.82	36.23	41.3 / 294.9
Energy (μJ)	-	-	-	144.5

---

[2]Some samples (with indices [18, 247, 493, 495, 717, 894, 904, 947] in test set) did not produce any spikes for an unknown reason. In that case, we removed the associated labels and compute the balanced accuracy on the 992 remaining samples.

Table 10.3   Perceived effort for each stage of the inference. 1: small, 5: large.

Board	A	S	G	M	I	Total
Kendryte	1	3	2	3	3	12
STM32L4R9	1	2	4	3	2	12
DynapCNN	3	1	3	1	1	9

10.4.2 Perceived Effort

Table 10.3 summarizes the team perceived effort for each of these phases in a qualitative manner. We observe a high variation in the effort perceived for each platform. The model preparation phase seems to be critical. In all the platforms, this phase is perceived as requiring a great effort. Kendryte K210 and STM32L4R9 require the most human intervention to build a complete deployment pipeline, while the deployment pipeline of DynapCNN is automated.

10.5 Conclusion

Although the development of embedded machine learning holds great promise, the lack of consistency and standardization across devices makes development extremely platform-dependent. Deploying a model on these devices requires to use of low-level tools, such as C language. However, most models are developed using (high-level) Python-based tools. The deployment process of a model therefore requires adaptation of the model from Python to C, which is time-consuming and is prone to errors and artifacts in the final implementation. Platform providers are aware of this problem and have started putting effort into facilitating the deployment by providing automated tools and interfaces with DL frameworks. Specifically, for the platforms used in these experiments, Sipeed has ported MicroPython to the Maix Dock, allowing to write code close to the one used to train the model; SynSense provides a library that allows interaction with the DynapCNN directly from a Python script, and allow simulation of the model before deployment, to get a quick idea of performance. Finally, the well-established ST-Microelectronic provides the X-CUBE-AI tool, which, in addition to analyzing the model before deployment, offers the possibility of validating the model on the target and retrieves relevant metrics without writing a single line of code.

However, these tools are recent and standards are not yet established. To promote and accelerate the development of machine learning on embedded interfaces, it is necessary to provide standardized tools accessible to model developers, where a minimum of knowledge about the platform is required. This will increase the adoption of the technologies. Some points seem essential to facilitate the adoption of low-power technologies, in particular:

- Up-to-date documentation: documents specifying platform schematics, APIs and dependencies on external tools must be carefully maintained.
- The documentation should contain examples for each API call.
- Model conversion tools should be compatible with most deep learning libraries (Tensorflow and PyTorch) and should detail which version and which operations (layers) are supported by each version of the tool. Ideally, conversion tools should be based on community standards, such as the ONNX format.
- Model conversion tools should be automated and provide understandable warnings and error messages.

To reduce the entry barrier for these low-power platforms for developers of Deep Learning models the following interfaces would be beneficial:

- A hardware simulation interface, in order to obtain a quick feedback on the feasibility of deploying the model on the platform, and to provide an interpretable error in case of memory exhaustion or unsupported layer.
- An evaluation of the key performance indicators relevant for edge computing, such as memory consumption, model speed (number of cycles per inference) and energy used during inference.

These interfaces will enable rapid prototyping and comparison of models for the Edge, while providing a solid foundation for iterating and developing new inference techniques.

Acknowledgements

This work is supported through the project ANDANTE. ANDANTE has received funding from the ECSEL Joint Undertaking (JU) under grant agreement No 876925. The JU receives support from the European Union's Horizon 2020 research and innovation programme and France, Belgium, Germany, Netherlands, Portugal, Spain, Switzerland. The authors are responsible for the content of this publication.

References

[1] Q. Liu, O. Richter, C. Nielsen, S. Sheik, G. Indiveri, and N. Qiao. Live demonstration: face recognition on an ultra-low power event-driven

convolutional neural network asic. In *Proceedings of the IEEE/CVF Conference on Computer Vision and Pattern Recognition Workshops*, pages 0–0, 2019.

[2] Canaan website. Kendryte K210 description page, 2022.

[3] C. R. Banbury, V. J. Reddi, M. Lam, W. Fu, A. Fazel, J. Holleman, X. Huang, R. Hurtado, D. Kanter, A. Lokhmotov, et al. Benchmarking tinyml systems: Challenges and direction. *arXiv preprint arXiv:2003.04821*, 2020.

[4] C. Ostrau, C. Klarhorst, M. Thies, and U. Rückert. Benchmarking of neuromorphic hardware systems. In *Proceedings of the Neuro-inspired Computational Elements Workshop*, pages 1–4, 2020.

[5] L. Heim, A. Biri, Z. Qu, and L. Thiele. Measuring what really matters: Optimizing neural networks for tinyml. *arXiv preprint arXiv:2104.10645*, 2021.

[6] V. Falbo, T. Apicella, D. Aurioso, L. Danese, F. Bellotti, R. Berta, and A. D. Gloria. Analyzing machine learning on mainstream microcontrollers. In *International Conference on Applications in Electronics Pervading Industry, Environment and Society*, pages 103–108. Springer, 2019.

[7] R. Sanchez-Iborra and A. F. Skarmeta. Tinyml-enabled frugal smart objects: Challenges and opportunities. *IEEE Circuits and Systems Magazine*, 20(3):4–18, 2020.

[8] A. Osman, U. Abid, L. Gemma, M. Perotto, and D. Brunelli. Tinyml platforms benchmarking. In *International Conference on Applications in Electronics Pervading Industry, Environment and Society*, pages 139–148. Springer, 2022.

[9] M. de Prado, M. Rusci, A. Capotondi, R. Donze, L. Benini, and N. Pazos. Robustifying the deployment of tinyml models for autonomous mini-vehicles. *Sensors*, 21(4):1339, 2021.

[10] Y. Lecun, L. Bottou, Y. Bengio, and P. Haffner. Gradient-based learning applied to document recognition. *Proceedings of the IEEE*, 86(11):2278–2324, 1998.

11

Efficient Edge Deployment Demonstrated on YOLOv5 and Coral Edge TPU

Ruben Prokscha, Mathias Schneider, and Alfred Höß

Ostbayerische Technische Hochschule Amberg-Weiden, Germany

Abstract

The recent advancements towards Artificial Intelligence (AI) at the edge resonate with an impression of a dichotomy between resource intensive, highly abstracted Machine Learning (ML) research and strongly optimized, low-level embedded design. Overcoming such opposing mindsets is imperative for enabling desirable future scenarios such as autonomous driving and smart cities. edge AI must incorporate both straightforward streamlined deployments together with resource efficient execution to achieve general acceptance. This research aims to exemplify how such an endeavour could be realized, utilizing a novel low power AI accelerator together with a state-of-the-art object detection algorithm. Different considerations regarding model structure and efficient hardware acceleration are presented for deploying Deep Learning (DL) applications in resource restricted environments while maintaining the comfort of operating at a high degree of abstraction. The goal is to demonstrate what is possible in the field of edge AI once software and hardware are optimally matched.

Keywords: edge AI, object detection, deep learning, YOLO, embedded systems, tensor processing unit.

11.1 Introduction

With AI shifting from a simple research subject towards end user applications, the issue of efficient deployment moves into focus. ML workloads

DOI: 10.1201/9781003377382-11

are decidedly different from average computing tasks. Hence, GPUs were the common solution for such undertakings. Realizing mobile intelligent appliances, requires even more specialized, low power accelerators which can be integrated into embedded environments. Such edge solutions attracted increasing interest within the last years. The European Strategic Research and Innovation Agenda (SRIA) [1] concretizes the term even further by introducing the terms Micro-, Deep- and Meta-edge. There are several different solutions available which target this new frontier. Most prominent are the NVIDIA Jetson family, which utilizes optimized embedded GPUs, the Intel Neural Compute Stick 2 which is comprised of a specialized Vision Processing Unit (VPU) and the Google Coral edge Tensor Processing Unit (TPU), which will be the focus of this work. As such, its impact on related research is presented in the following section. The task of object detection was chosen to be part of the experimental test setup for evaluating the accelerator. You Only Look Once (YOLO) version 5 [2] serves as delegate for these class of networks in the upcoming section. It is evaluated, how models can be modified to facilitate edge TPU characteristics. Furthermore, it is shown how this optimized solution compares to models provided by Google. With a focus on deployment, a lightweight software stack is introduced which enables efficient AI solutions without sacrificing high-level development. Finally, a conclusion is provided giving a synapsis of the key findings and offering points of interest for future work.

11.2 Related Work

In recent years, the usage of decentralized AI at the edge has become a progressively relevant research topic. Thereby, besides GPU acceleration, the energy-efficient edge TPU was of special interest by research fellows. For applications with strict power or battery limitation, such as in the area of UAV, the usage of the edge TPU is evaluated in recent work. Thereby, applications comprise indoor person-following systems [3], vision-based trash and litter detection [4], and lightweight odometry estimation [5]. Using a U-Net network architecture, Roesler et al. leverage their edge AI setup combining the edge accelerator with a STM32MP157C-DK2 board for the yield estimation of grapes in an agriculture use case [6]. But also, other application domains are explored, e.g., in [7], which utilizes the edge TPU to process time-series data to determine the remaining useful life. Since at that time Recurrent Neural Networks (RNNs) were not yet supported by the accelerator, their model architecture employs a deep Convolutional Neural Network (CNN). It is worth mentioning that their experiments included measurements for models using quantization-aware training as well as post-training quantization, which outperformed reference CPU and GPU deployments in terms of latency and accuracy. The authors in [8] examine the potential of the edge TPU for detecting network intrusion to ensure security at the edge using feed forward and CNN architectures. They elaborate their classification scores on a public benchmark dataset, and further investigate the energy efficiency of their DL algorithms in comparison to traditional CPU processing. Their studies on the effects of larger model sizes reveal a bimodal behaviour of the edge accelerator, indicating a decline of the energy efficiency ratio as soon as a certain model size is exceeded. This finding is the focus of their consecutive work and is confirmed by more refined experiments [9].

Besides this applied research of utilizing the edge accelerator for a dedicated application, more theoretical research was conducted to explore and demarcate TPU capabilities. Therefore, several benchmarks were performed to determine its performance empirically using various setups differing in the models under test, obtained metrics, or compared edge devices [10, 11, 12]. Providing micro-architectural insights, Google researcher, Yazdanbakhsh et al., elaborate an extensive evaluation covering different structures in CNNs and their effects on latency and energy consumption [13]. With a similar level of hardware details, the authors in [14] analysed the inference of 24 Google edge models, revealing major shortcomings of the edge TPU architecture which must be taken into account for efficient deployment. Furthermore, they incorporate the results into their framework for heterogeneous edge ML accelerators called Mensa, improving the edge TPU performance significantly.

11.3 Experimental Setup

Figure 11.1 depicts the setup used for this research. A Raspberry Pi 4 Model B with 4 GB memory served as base platform. The Google Coral edge TPU accelerator was connected either to a USB2 or USB3 port for performance and accuracy evaluation.

11.3.1 Google Coral Edge TPU

Google developed a custom Application Specific Integrated Circuit (ASIC) for edge inference. This specialized TPU can be connected to existing systems utilizing a USB, (m)PCIe or M.2 interface. Figure 11.1 depicts the USB

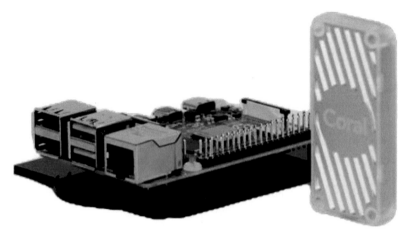

Figure 11.1   Raspberry Pi 4 with Google Coral edge TPU USB accelerator.

dongle variant of the accelerator with is advertised to perform up to four trillion operations per second. Approximately 8 MB 'scratchpad' memory is available per unit and the peak power consumption is rated at 2 W [15]. Additionally, multiple of these coprocessors can be chained together for handling bigger workloads. The TPU hardware operates on 8 Bit integer variables. Both performance and power consumption benefit from a reduced complexity in the hardware design. However, this introduces weight quantization as an additional step before deployment. The reduction in precision from floating point to 8 Bit integers variables subsequently leads to a deterioration of accuracy. Further overhead is introduced by the addition of quantization operations to the execution graph.

Deploying a model for this device entails several pitfalls due to a rather convoluted development pipeline. Google necessitates its own Tensorflow (TF) framework as starting point. Hence, models from other frameworks must be converted by means of e.g., Open Neural Network Exchange (ONNX). There, a quantization step is performed alongside a conversion to the TFLite format. The final step involves a proprietary edge TPU compiler, which translate the TFLite instructions for the edge TPU. Inference on the other hand is straight forward. The TFLite runtime provides the interfaces for loading and executing the model file, while the libedgetpu is responsible for handling the low-level communication with the accelerator. This allows for a very lightweight deployment of 10 MB to 20 MB (without model) compared to conventional GPU solutions, which can require over a gigabyte disk storage for the libraries alone.

11.3.2 YOLOv5

The original You Only Look Once (YOLO) architecture was proposed by Joseph Redmon in 2016 [16]. It performs both object detection and classification in a single model. This resulted in a significant performance increase compared to classical two stage designs (e.g., Region Based Convolutional Neural Networks (R-CNNs) [17]). Since the original design, many improvements were made. YOLOv5 [2] is based on the YOLOv3 [18] architecture. It is under constant open-source development by Ultralytics, who shifted the focus from academic research to accessible deployment. They provide an end-to-end solution which allows for training, testing and exporting models to a variety of different deployment frameworks. This includes the integration of the previously described pipeline for generating edge TPU models from version 6.1 onward.

11.4 Performance Considerations

The Coral accelerator achieves its low energy footprint and high performance by sacrificing flexibility. This manifests itself in a significantly reduced instruction set [19]. The edge TPU compiler is a black box which aims to aggregate as much operations as possible and convert them into a binary which can be executed by the coprocessor. Every operation, which is not mapped accordingly, must therefore run-on CPU. This section aims to provide guidance for optimizing a model for edge TPU execution exemplified on YOLOv5 (release 6.1).

11.4.1 Graph Optimization

Figure 11.2 depicts the graphs of two edge TPU models. Figure 11.2a shows the small variant of the YOLOv5 model with additional optimizations. The EfficientDet Lite0 [20] model in Figure 11.2b was taken from the Coral model zoo [21]. Most of the graph is mapped to the edgetpu-custom-op, while some operations are still executed by the main processor. In the following, possible issues are shown when compiling a model and ways to improve the mapping are elaborated.

11.4.1.1 Incompatible Operations

The compiler only maps operations until it encounters an incompatibility. Everything after that is executed on the CPU. This is especially critical for activation functions (e.g., LeakyReLU, Hardswish), as they are distributed

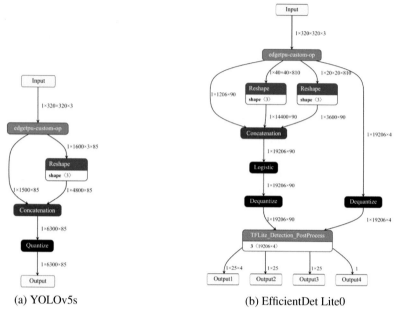

(a) YOLOv5s   (b) EfficientDet Lite0

Figure 11.2   Quantized edge TPU Models.

throughout the graph. While it is possible to create multiple TPU subgraphs, the overhead of transferring intermediate tensors several times between CPU and TPU usually eliminates any benefits. It is therefore advisable to use compatible activation functions (e.g., ReLU, Logistic). Furthermore, binary operations (e.g., AND, OR) are also not supported.

11.4.1.2 Tensor Transformations

The reshape and transpose operation are not mapped once their input tensor exceeds a certain soft threshold. There is no documentation on how this limit is calculated and it seems to be dependent on the general model structure. However, it could be observed, that this threshold is significantly smaller for the transpose operation. A possible explanation for this behaviour could be an inability of the accelerator to address memory in a different order. A transpose operation on CPU would imply a change in the direction (column/row wise) memory is read from the same location. If this is not supported by the TPU, memory reallocation is required.

There are several ways for addressing this issue. One approach is to reduce the size of the input tensor. In CNNs the size is proportionally

Table 11.1 Comparison of YOLOv5s model before and after optimizations.

Input Size	YOLOv5s (6.1)		YOLOv5s (6.2dev)		Speedup
	TPU/CPU	**USB3 Speed**	**TPU/CPU**	**USB3 Speed**	
320x320	245/40	30.10ms	253/3	24.27ms	19.37%
640x640	225/59	427.48ms	240/16	178.67ms	58.20%

propagated through the network. Hence, reducing the input size results in smaller intermediate tensors. Further reduction can be induced by limiting the number of output classes. If graph modifications are viable, a divide and conquer strategy can be used to split tensors before the operation and merging afterwards. Moving these operations to the bottom of the graph can also be an option as the instruction are fast on CPU. A last option is using mathematical transformation to change the graph beneficially.

Some of these strategies were used to optimize the YOLOv5 models which are evaluated further in this research. All changes were committed to the open-source project in a pull request [22] and are part of the next major release (6.2). Table 11.1 shows the performance impact for the demonstrator setup. Both model variants experienced a significant speedup in inference time. The variant with the larger input size improves significantly.

11.4.2 Performance Evaluation

In the following, different variants of the optimized YOLOv5 models are compared to other object detectors supplied by Google. All numerical values can be found in Table 11.2. The inference speed was evaluated utilizing the Google benchmark model tool [23]. Version 16 of libedgetpu-max was used, and each inference was repeated 100 times with a previous warm-up phase. Accuracy was determined by pycocotools and the Common Objects in Context (COCO) evaluation dataset [24]. The input images were proportionally scaled to input size with bilinear interpolation. The Google models have a postprocessing operation integrated in the model graph (ref. Figure 11.2b). It was evaluated separately for inference speed and fast Non-Maximum Suppression (NMS) [25] was used for all models as it is the default setting of this custom operation. Furthermore, the threshold for confidence was set to 0.001 and overlap to 0.65.

11.4.2.1 Speed-Accuracy Comparison

Figure 11.3 shows the mean average precision ($mAP_{50:95}$) of each tested model in relation to the inference speed. It can be observed that the edge TPU

Table 11.2   Model comparison in regards of input size, file size, operation

	Input Size	Size in MB	Ops		Accuracy						Speed in ms			
			TPU/CPU	mAP_{50-95}	mAP_{50}	mAP_{75}	mAP_S	mAP_M	mAP_L	USB2		USB3	
EfficientDet Lite0	320x320	5.93	260/7	24.30	39.10	25.40	5.50	27.30	42.20	164.06	+21.6	57.57	+21.79
EfficientDet Lite1	384x384	8.01	315/7	28.90	45.10	30.70	8.80	33.20	47.00	243.16	+30.57	81.43	+31.03
EfficientDet Lite2	448x448	10.67	349/8	32.30	48.90	34.80	12.20	37.10	49.60	490.06	+40.44	152.77	+39.58
SSD Mobilenetv2	300x300	7.08	99/3	15.50	27.20	15.30	1.20	10.90	34.30	31.68	+2.01	10.90	+2.06
SSDLite MobileDet	320x320	5.4	134/3	22.50	38.30	23.30	2.50	19.40	48.20	34.98	+8.77	9.56	+5.93
YOLOv5n	320x320	2.38	253/3	18.10	32.40	18.10	3.00	17.60	32.60	64.35		24.64	
	480x480	2.47	240/16	22.20	39.60	22.50	6.10	24.60	34.40	127.57		50.41	
	640x640	2.43	240/16	22.70	41.10	22.80	8.90	26.70	31.00	262.79		93.60	
	320x320	7.84	253/3	26.10	43.50	27.00	6.40	27.80	45.10	67.76		24.27	
YOLOv5s	400x400	7.96	240/16	28.50	47.40	29.40	8.90	31.60	45.10	162.15		49.42	
	480x480	8.09	240/16	29.80	49.20	31.30	11.30	33.70	44.70	238.43		77.77	
	640x640	8.78	240/16	30.20	50.50	31.60	14.10	35.00	41.10	711.97		178.67	
	240x240	22.13	325/3	28.80	46.70	30.30	6.80	31.00	51.10	460.60		63.61	
YOLOv5m	320x320	22.32	325/3	32.40	51.30	34.40	10.40	36.10	53.60	542.08		85.98	
YOLOv5l	480x480	23.33	313/16	36.90	58.00	39.10	16.20	41.90	53.30	1268.55		238.90	
	320x320	47.77	399/3	35.40	55.10	37.60	12.70	40.00	56.60	1296.39		177.60	

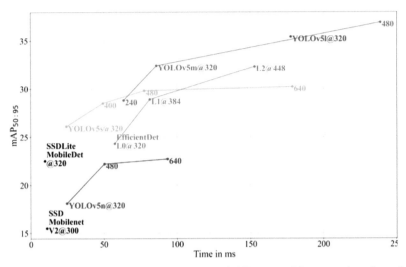

Figure 11.3 USB3 speed-accuracy comparison of different model types and configurations for edge TPU deployment.

works best lower input sizes, while larger inputs cause an unproportionate slowdown compared to the benefit in accuracy. Interesting are the nano and small models with 320 px input. They have an almost identical inference time, while the accuracy of the s-model is significantly better. They share the same vertical graph structure, while the larger one is scaled horizontally by a factor of two. Hence, the small variant has twice as many weights for each convolutional layer. This aligns with the insights from [12] that horizontal scaling is preferable. The model should be very close to a sweet spot, for which all weights are cached within the 8 MB device memory. Sacrificing some model vertical space for more width could theoretically improve the accuracy even further.

In general, YOLOv5 performs better than the other models. Only the nano model has issues, which is probably caused by its particularly small file size. If speed is the deciding factor, SSDLite MobileDet [26] [27], is still the preferable solution. The classical SSD Mobilenetv2 [26] [28] does not seem to be competitive anymore. The EfficientDet models perform reasonable, however considering the additional overhead by a particularly slow postprocessing operation, YOLOv5 should be considered the better solution. All models share a low accuracy for small objects, which could be an issue inflicted by quantization.

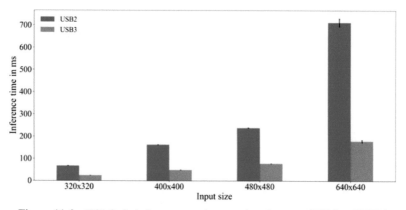

Figure 11.4   YOLOv5s inference speed comparison between USB2 and USB3

11.4.2.2 USB Speed Comparison

Considering the purpose of edge accelerators to allow for AI deployment on low power devices, USB3 might not always be an option. Hence, it should be evaluated whether a deployment utilizing USB2 is a viable option. The maximum speed for such a connection is rated at 60 MB/s, while USB3 is specified at almost ten times this value.

Figure 11.4 depicts the speed comparison the small model with varying input size. A considerable difference for the inference speed can be observed. The USB2 interface causes a slowdown by a factor of three. The model parameters should fit entirely into the device memory. Therefore, only the data transfer should impact the speed. Equation (11.1) shows how the data flowing to and from the device is calculated. $data_{in}$ only depends on the input size, while $data_{out}$ also considers the number of anchor boxes (3), strides for multi scale outputs (8, 16, 32) and class count. For the 320 px model, this results in 842.7 KB of data flow per inference, while the 640 px input increases this value to 3.37 MB. Additional data flow could arise due to intermediate tensors, which are too large to be buffered on the device. Whether this is an issue here must be determined in future research.

$$
\begin{aligned}
data_{in} &= 3x_{in}y_{in} \\
data_{out} &= 3\left(\frac{1}{8^2} + \frac{1}{16^2} + \frac{1}{32^2}\right)x_{in}y_{in} * (5 + n_{cls})
\end{aligned} \tag{11.1}
$$

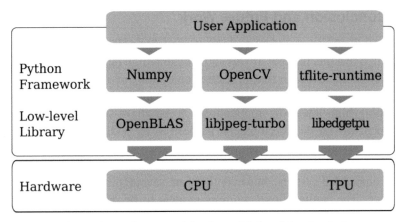

Figure 11.5 Micro software stack for fast and lightweight edge deployment.

11.4.3 Deployment Pipeline

An AI application can be considered a data pipeline of the steps. At first data must be loaded and pre-processed to comply with the model. In the context of object detection, this implies loading and scaling a jpeg image. The following steps are inference and postprocessing. The latter takes the raw model output and transforms into a usable form. This could involve thresholding, NMS and coordination transforms. The pipeline is executed for each inference, hence all steps should be highly optimized. Most efforts are usually focused towards optimizing the model while neglecting everything else. This section introduces a small deployment stack for object detection, which is both optimized and allows for the usage of well-established high-level frameworks.

The software stack depicted in Figure 11.5 shows a simple layer model for a lightweight deep vision deployment. The part concerning the TPU was previously elaborated. Loading and transforming images is often handled by OpenCV [29]. It uses shared low-level libraries to perform these operations. Providing an optimized image loader, such as libjpeg-turbo [30] can therefore accelerate the whole pipeline. Similar is true for Numpy [31], which is responsible for performing mathematical tensor operations on CPU. A dedicated math library such as OpenBLAS [32] makes use of Single Instruction Multiple Data (SIMD) which performs vector operations faster and more efficient. Such a software stack is similarly fast compared to a solution written in a compiled language, while being way more flexible. It could also be viable to package such an application into a lightweight container for easy deployment using virtualization technologies.

11.5 Conclusion and Future Work

This research demonstrated how efficient edge AI applications can be implemented in a feasible manner. It was shown that a high degree of optimization is required to make the best use of limited computing resources. Additionally, a lightweight software stack was presented, which can be used as basis for building high level ML applications. A paradigm shift towards a more deployment driven AI development, as portrait by YOLOv5, is mandatory for making ubiquitous AI possible. The Google Coral edge TPU offers high potential for enabling real-time object detection for common video stream rates on embedded systems, however there are several pitfalls associated with the device. The limited opset requires models to be designed accordingly, which must be in the interest of the developers. Another issue is the USB2 performance. Future research must evaluate, what exactly causes this drastic slowdown. If the TPU should be used in ultra-low power segments (e.g., Micro Controller Units), USB3 will not be viable. Changing the model to reduce the amount of data flowing to and from the device could alleviate this shortcoming.

Acknowledgements

This work has been financially supported by the AI4DI project. AI4DI receives funding within the Electronic Components and Systems For European Leadership Joint Undertaking (ESCEL JU) in collaboration with the European Union's Horizon 2020 Framework Programme and National Authorities, under grant agreement n° 826060.

References

[1] AENEAS, Inside Industry Association, and EPOSS. ECS – Strategic Research and Innovation Agenda 2022. en. Jan. 2022. URL: https://ecscollaborationtool.eu/publication/download/slides-ovidiu-vermesan.pdf (visited on 03/31/2022).

[2] G. Jocher et al. ultralytics/yolov5: v6.1 - TensorRT, TensorFlow Edge TPU and OpenVINO Export and Inference. Feb. 2022. URL: https://zenodo.org/record/6222936 (visited on 03/30/2022).

[3] A. Boschi et al. "A Cost-Effective Person-Following System for Assistive Unmanned Vehicles with Deep Learning at the Edge". en. In: Machines 8.3 (Aug. 2020), p. 49.

[4] M. Kraft et al. "Autonomous, Onboard Vision-Based Trash and Litter Detection in Low Altitude Aerial Images Collected by an Unmanned Aerial Vehicle". en. In: Remote Sensing 13.5 (Mar. 2021), p. 965.

[5] N. J. Sanket et al. "PRGFlow: Benchmarking SWAP-Aware Unified Deep Visual Inertial Odometry". en. In: arXiv:2006.06753 [cs] (June 2020).

[6] M. Roesler et al. "Deploying Deep Neural Networks on Edge Devices for Grape Segmentation". en. In: Smart and Sustainable Agriculture. Ed. by Selma Boumerdassi, Mounir Ghogho, and Éric Renault. Vol. 1470. Cham: Springer International Publishing, 2021, pp. 30–43.

[7] C. Resende et al. "TIP4.0: Industrial Internet of Things Platform for Predictive Maintenance". en. In: Sensors 21.14 (July 2021), p. 4676.

[8] S. Hosseininoorbin et al. "Exploring Edge TPU for Network Intrusion Detection in IoT". en. In: arXiv:2103.16295 [cs] (Mar. 2021).

[9] S. Hosseininoorbin et al. "Exploring Deep Neural Networks on Edge TPU". en. In: arXiv:2110.08826 [cs] (Oct. 2021).

[10] M. Alnemari and N. Bagherzadeh. "Efficient Deep Neural Networks for Edge Computing". en. In: 2019 IEEE International Conference on Edge Computing (EDGE). Milan, Italy: IEEE, July 2019, pp. 1–7.

[11] M. Antonini et al. "Resource Characterisation of Personal-Scale Sensing Models on Edge Accelerators". en. In: Proceedings of the First International Workshop on Challenges in Artificial Intelligence and Machine Learning for Internet of Things. New York NY USA: ACM, Nov. 2019, pp. 49–55.

[12] A. A. Asyraaf Jainuddin et al. "Performance Analysis of Deep Neural Networks for Object Classification with Edge TPU". In: 2020 8th International Conference on Information Technology and Multimedia (ICIMU). Aug. 2020, pp. 323–328.

[13] A. Yazdanbakhsh et al. An Evaluation of Edge TPU Accelerators for Convolutional Neural Networks. Feb. 2021.

[14] A. Boroumand et al. "Google Neural Network Models for Edge Devices: Analyzing and Mitigating Machine Learning Inference Bottlenecks". en. In: arXiv:2109.14320 [cs] (Sept. 2021).

[15] USB Accelerator datasheet. en-us. URL : https://coral.ai/docs/accelerator/datasheet/ (visited on 03/31/2022).

[16] J. Redmon et al. "You Only Look Once: Unified, Real-Time Object Detection". In: 2016 IEEE Conference on Computer Vision and Pattern Recognition (CVPR). June 2016, pp. 779–788.

[17] R. Girshick et al. "Rich Feature Hierarchies for Accurate Object Detection and Semantic Segmentation". In: 2014 IEEE Conference on Computer Vision and Pattern Recognition. June 2014, pp. 580–587.

[18] J. Redmon and A. Farhadi. "YOLOv3: An Incremental Improvement". In: (Apr. 2018).

[19] TensorFlow models on the Edge TPU. en-us. URL: https://coral.ai / docs / edgetpu / models - intro / #supported – operations (visited on 03/30/2022).

[20] M. Tan, R. Pang, and Q. V. Le. "EfficientDet: Scalable and Efficient Object Detection". In: arXiv:1911.09070 [cs, eess] (July 2020). arXiv: 1911.09070.

[21] Models - Object Detection. en-us. URL: https://coral.ai/models/object-detection/.

[22] EdgeTPU optimizations by paradigmn Pull Request #6808 ultra-lytics/yolo5. en. URL: https://github.com/ultralytics/yolov5/pull/6808 (visited on 03/31/2022).

[23] Performance measurement — TensorFlow Lite. en. URL : https://www.tensorflow.org/lite/performance/measurement (visited on 03/30/2022).

[24] T.-Y. Lin et al. "Microsoft COCO: Common Objects in Context". en. In: Computer Vision – ECCV 2014. Ed. by David Fleet et al. LectureNotes in Computer Science. Cham: Springer International Publishing, 2014, pp. 740–755.

[25] J. Hosang, R. Benenson, and B. Schiele. "Learning Non-maximum Suppression". In: 2017 IEEE Conference on Computer Vision and Pattern Recognition (CVPR). ISSN: 1063-6919. July 2017, pp. 6469–6477.

[26] W. L. et al. "SSD: Single Shot MultiBox Detector". en. In: Computer Vision – ECCV 2016. Ed. by Bastian Leibe et al. Lecture Notes in Computer Science. Cham: Springer International Publishing, 2016, pp. 21–37.

[27] Y. Xiong et al. "MobileDets: Searching for Object Detection Architectures for Mobile Accelerators". In: arXiv:2004.14525 [cs] (July 2020). arXiv: 2004.14525.

[28] M. Sandler et al. "MobileNetV2: Inverted Residuals and Linear Bottlenecks". In: 2018 IEEE/CVF Conference on Computer Vision and Pattern Recognition. June 2018, pp. 4510–4520.

[29] G. Bradski. "The OpenCV Library". In: Dr. Dobb's Journal of Software Tools (2000).

[30] libjpeg-turbo. original-date: 2015-07-27T07:11:54Z. Mar. 2022. URL: https://github.com/libjpeg-turbo/libjpeg-turbo (visited on 03/31/2022).

[31] C. R. Harris et al. "Array programming with NumPy". en. In: Nature 585.7825 (Sept. 2020), pp. 357–362.

[32] Q. Wang et al. "AUGEM: automatically generate high performance dense linear algebra kernels on x86 CPUs". en. In: Proceedings of the International Conference on High Performance Computing, Networking, Storage and Analysis. Denver Colorado: ACM, Nov. 2013, pp. 1–12.

12

Embedded Edge Intelligent Processing for End-To-End Predictive Maintenance in Industrial Applications

Ovidiu Vermesan[1] and Marcello Coppola[2]

[1]SINTEF AS, Norway
[2]STMicroelectronics, France

Abstract

This article advances innovative approaches to the design and implementation of an embedded intelligent system for predictive maintenance (PdM) in industrial applications. It is based on the integration of advanced artificial intelligence (AI) techniques into micro-edge Industrial Internet of Things (IIoT) devices running on Arm® Cortex® microcontrollers (MCUs) and addresses the impact of a) adapting to the constraints of MCUs, b) analysing sensor patterns in the time and frequency domain and c) optimising the AI model architecture and hyperparameter tuning, stressing that hardware–software co-exploration is the key ingredient to converting micro-edge IIoT devices into intelligent PdM systems. Moreover, this article highlights the importance of end-to-end AI development solutions by employing existing frameworks and inference engines that permit the integration of complex AI mechanisms within MCUs, such as NanoEdge™ AI Studio, Edge Impulse and STM32 Cube.AI. Both quantitative and qualitative insights are presented in complementary workflows with different design and learning components, as well as in the backend flow for deployment onto IIoT devices with a common inference platform based on Arm® Cortex®-M-based MCUs. The use case is an n-class classification based on the vibration of generic motor rotating equipment. The results have been used to lay down the foundation

DOI: 10.1201/9781003377382-12

of the PdM strategy, which will be included in future work insights derived from anomaly detection, regression and forecasting applications.

Keywords: predictive maintenance, smart sensors systems, industrial internet of things, industrial internet of intelligent things, vibration analysis, machine learning, deep learning architecture, edge-embedded devices.

12.1 Introduction and Background

Leveraging AI methods and techniques at the edge is vital for increasing the performance and capabilities of the intelligent sensor systems and IIoT devices used in industrial manufacturing. For many intelligent applications, the edge AI processing concept is reflected in the emergence of different edge layers (micro-, deep-, meta-edge). The edge processing continuum includes the sensing, processing and communication devices (micro-edge) close to the physical industrial assets under monitoring, the gateways and intelligent controllers processing devices (deep-edge), and the on-premise multi-use computing devices (meta-edge). This continuum creates a multi-level structure that moves up in processing, intelligence, and connectivity capability.

Micro-edge devices are typically small sensors and actuators equipped with microcontrollers (MCUs) based on Arm® Cortex®-M cores (e.g., M0, M0+, M3, M4, M7) or open-source RISC-V instruction set architecture, circuits with memory, serial ports, peripherals, and wireless capabilities and designed to perform and extend the specific tasks of embedded systems.

Developing AI functionalities for micro-edge devices is a complex process that has increased potential in various industrial applications, including manufacturing. In industrial manufacturing, the implementation of machine learning (ML) and deep learning (DL) models on micro-edge-embedded devices has an absolute advantage for condition monitoring and PdM/prescriptive maintenance (PsM) operations for industrial motors/equipment. Using AI-enabled micro-edge devices for motors/equipment monitoring in industrial processes can prevent downtime by alerting users to perform preventative maintenance based on equipment real-time conditions.

There are several works that provide a comprehensive review of frameworks available in the market that currently permit the integration of complex ML and DL mechanisms within MCUs [1][4].

This article researches and investigates different approaches to using ML and DL technologies to bring AI capabilities to micro-edge devices and applies these capabilities for classification for PdM industrial applications. The goal is to implement ML and DL techniques in low-energy systems, including sensors, to perform intelligent automated tasks, such as PdM and PsM.

The approaches used in this article illustrate how to optimise ML and DL models for resource-constrained micro-edge-embedded devices. The article gives an overview of the data acquisition and prediction aspects of ML and DL, discusses how to build ML and DL models targeting micro-edge devices and presents the experimental results using different tools and approaches.

The article is organised into five sections. The introduction on intelligent edge processing real-time maintenance systems and description of data-, model- and knowledge-driven methods for time series is included in Section 12.1. Section 12.2 describes the architecture and design of motor classification for PdM, including methods and possible end-to-end flows and presents the use case, i.e., motor classification. Section 12.3 introduces the implementation of the classification use case using three existing platforms. Section 12.4 highlights specific experiments performed and the results that were achieved through the lens of employing different tools. Section 12.5 addresses future research challenges and discusses the key open issues related to AI techniques and methods in implementing intelligent edge processing real-time maintenance systems for the purposes of industrial applications.

12.2 Machine and Deep Learning for Embedded Edge Predictive Maintenance

For industrial manufacturing facilities using motors in the process line, the maturity of maintenance practices is a crucial determinant of the ability to operate reliably and profitably without interruption. Condition-based monitoring maintenance (CBM) addresses uptime and maintenance costs by monitoring one or several critical measurements for the motors, such as temperature, vibration, oil analysis and current, which are used as indicators of an out-of-specification condition. Maintenance tasks are performed when needed. PdM applies a more extensive set of input data and more analysis to provide a more reliable indicator of the overall health and condition of the motor as well as a more accurate prediction of a possible failure and what action should be considered to prevent it.

With PdM, the motors are serviced considering the actual wear and tear and service needs, reducing unexpected outages, making fewer scheduled maintenance repairs or replacements, and using fewer maintenance resources (including spare parts and supplies) while simultaneously decreasing failures. PdM provides the prerequisite foundation for PsM and autonomous maintenance (by executing actions automatically, without human intervention). PsM builds on the infrastructure and data collected for PdM, following the various corrective actions taken by maintenance personnel and the resulting outcomes.

Figure 12.1 illustrates a typical industrial motor with a rotor, stator, bearings, and shaft as essential components for the engine's normal operation.

The various components conditions and operations are possible causes that can generate anomalous behaviour, thus defining various abnormal states (classes). A large amount of historical and real-time information is required to identify, classify, and predict motor's possible failures. AI-based ML and DL algorithms are suitable to deal with these types of tasks.

This paper focuses on AI-based PdM approaches, which learn from historical and real-time data and recommend the best timing and course of action for a given set of conditions and sub conditions employing ML and DL models implemented using micro-edge-embedded devices. For example, the implementation of an ML solution into a PdM application includes several steps: data preparation, feature engineering, algorithm selection and parameter tuning.

The interaction between the edge IIoT devices, ML and DL have opened opportunities for new data-driven approaches for PdM solutions in industrial processes. In this paper, different techniques and tools were successfully tested using various methods based on ML and DL to predict the state of industrial motors and to detect and classify motors conditions based on trained data. The PdM monitoring has been tested on measurements

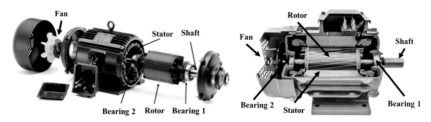

Figure 12.1   Industrial motor components [5] [6]

performed on bench motors using computation at the micro-edge, allowing real-time acquisition, processing, and wireless communication.

12.3 Approaches for Predictive Maintenance

AI-based PdM approaches [2][3][7], employing ML and DL models implemented using micro-edge-embedded devices, are designed on different hardware platforms and software suites, generating embedded code, and performing learning and inference engine optimisations. Depending on the application and the frameworks and inference engines for integrating AI mechanisms within MCUs, several variants of the workflows are used.

This paper focuses on NanoEdge^{TM} AI (NEAI) Studio [14], Edge Impulse (EI) [8][10] and STM32 Cube.AI [10][13].

Table 12.1 gives an overview of the features of these frameworks, which support the workflows of ML and DL model development and deployment on microcontroller class devices. AI/ ML models work on frameworks such as Keras, ONNX, Lasagne, Caffe, Convetjs etc.

Table 12.1 Frameworks and inference engines for integrating AI mechanisms within MCUs

Framework	Platforms	Models	Training Libraries
Edge Impulse (EI)	Arm® Cortex®-M, TI CC1352P, Arm® Cortex®-M -A, Espressif ESP32, Himax WE-I Plus, TENSAI SoC	NN, k-means, regressors (including feature extraction)	Tensor Flow, Scikit-Learn
Nano Edge AI Studio (NEAI)	Arm® Cortex®-M (STM32 series)	Unsupervised learning	-
STM32Cube.AI	Arm® Cortex®-M (STM32 series)	NN, k-means, SVM, RF, kNN, DT, NB, regressors	PyTorch, Scikit-Learn, Tensor Flow, Keras, Caffe, MATLAB, Microsoft Cognitive Toolkit, Lasagne, ConvNetJS

12.3.1 Hardware and Software Platforms

The experiments in this paper perform the processing of various types of input data, including three-axis vibration, temperature, and device logs. The data for the experiments was collected from bench motors using a STWIN Sensor Tile Wireless Industrial Node IIoT device.

This micro-edge IIoT device comprises of three axis ultrawide bandwidth (DC to 6 kHz) acceleration sensor (ISM330DHCX), a 12-bit analog-to-digital converter, a user-configurable digital filter chain, a temperature sensor, and a serial peripheral interface. The micro electro mechanical systems (MEMS) vibration sensor has a selectable sensitivity (± 2, ± 4, ± 8, or ± 16 g) and processing capabilities ensured by an Arm® Cortex®-M4 processor (120 MHz, 640 KB RAM, 2 MB Flash). The micro-edge device can be powered externally or by an internal lithium-ion battery and has BLE and Wi-Fi connectivity.

The design flow allows collecting or uploading training data from micro-edge devices, labelling the data, training an ML model, and launching and monitoring ML models in a production environment.

The PdM AI-based design flow uses the sensors and hardware platforms, software development kits (SDKs), frameworks and inference engines for integrating AI mechanisms within MCUs to generate code to be deployed on MCUs that allow running AI models in embedded systems by performing predictions at the edge. The ML and DL models deployed on the micro-edge devices become part of the firmware flashed into the MCUs.

A micro-edge AI processing flow is illustrated in Figure 12.2.

The AI-based flow uses an embedded compiler that can convert models to C/C++ to increase the efficiency of models trained on the edge platform and reduce RAM, storage usage and code size by tens of percent.

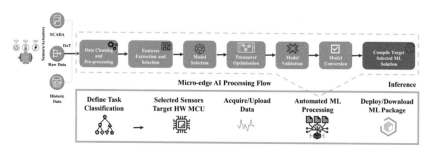

Figure 12.2   Micro-edge AI processing flow

12.3.2 Motor Classification Use Case

The use case analysed in this article is the classification of the state of a motor based on the vibration measurements using an accelerometer sensor from an IIoT device. The signals covering all states to be classified were collected using a built-in three-axis accelerometer (ISM330DHCX) to measure the accelerations of three orthogonal directions.

In general, the n-class classification of n different states uses static models with pretrained libraries.

The classes were defined based on conditions (motor speeds) and sub-conditions (malfunctions). The motor was operating at fixed speeds, which were divided into three classes based on various percentages of the maximum speed (50%, 75% and 100%). A malfunction of the motor (motor fan trepidations) was added to the second class to obtain a new class. The classes defined are:

- MOTOR_OFF: just record signals when nothing is happening
- MOTOR_ON_NORMAL_50: the motor is running at 50% of the maximum speed
- MOTOR_ON_NORMAL_75: the motor is running at 75% of the maximum speed
- MOTOR_ON_NORMAL_75_B: the motor fan produces additional trepidations to the motor, while the motor is running at 75% of the maximum speed
- MOTOR_ON_NORMAL_MAX: motor is running at maximum speed.

12.4 Experimental Setup

The design and implementation steps and the experimental setup of the end-to-end (E2E) classification application use two main primary flows, including NEAI Studio and EI. The former creates ML static libraries based on unsupervised algorithms, while the later employs deep neural networks (NNs) for the classification task. A third flow was branched out from EI into Python using Tensor Flow's Keras API, and the resulted model was fed onto STM32Cube.AI.

The experimental process started by collecting the vibration signals from the micro-edge IIoT device mounted on the motor, through a simple data-logger application in real-time. The recorded signals were then analysed in both the time and frequency domain, filtrated, and datasets were prepared for each flow. The classification AI models were then built in each flow, using the accelerometer spectral features (e.g., root mean square (RMS), frequency and amplitude of spectral power peaks, etc.) and optimise the performance. In the end the three models were deployed and integrated with the firmware using STM32 CubeIDE. Finally, inference classifications were run to note the performance of the implementations and deployments.

12.4.1 Signal Data Acquisition and Pre-processing

Prior to acquiring the signals, a thorough analysis of the vibration patterns of the motor have been conducted, landing to the conclusion that the most suitable sampling frequency to capture vibration patterns is 1667 Hz.

Both NEAI and EI offer several ways to take the measurements from the sensor IIoT device directly from within their GUIs. Acquiring signals with datalogger functionality in NEAI seemed to be the most straightforward data acquisition approach as it only requires the SD card. In the experimental use case, a simple logger application was used that reads and logs the raw accelerometer sensor data directly on the serial port, so that logs can be retrieved from a computer using serial tools such as Tera Term or from the console of the integrated development environment (IDE).

For the three-axis accelerometer sensor, a collection of signals (split in 60% training, 20% validation and 20% test) was acquired for each of the classes, with a buffer size of 512 samples on each axis, in total 1536 values per signal. Thus, with a sampling frequency of 1667 Hz, each buffer represents a snapshot of approximately 300 milliseconds of the accelerometer temporal vibration data, which is sufficient to capture the essence of the motor vibration patterns. The vibration signals collected are visualised as shown in Figure 12.3, in both temporal and frequency plots for the accelerometer sensor Z-axis for each of the two classes.

To be able to better differentiate the individual classes and thus ensuring high accuracy score, the recorded signals were processed in frequency domain. Filter settings was activated in the signals pre-processing steps. By providing filtering, only the frequencies that represent the characteristics of the motor vibration are kept, and the rest are attenuated. Filtering techniques also help to eliminate high frequency noise that interfere with the vibration signal, and eliminate frequencies for transitions between states, which would normally yield unknown class.

The recorded signals for each class were downloaded and then converted into a format accepted by EI, to ensure the same signals are being used for the signal processing, thus yielding similar results.

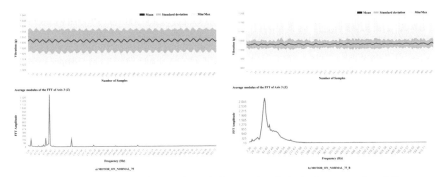

Figure 12.3 Visualisation of two selected classes signals in both temporal and frequency domain with NEAI

Till acceptable quality-labelled data sets were arrived at, several iterations were performed, and this included recording new signals without background noise, collecting/recording longer signals and even changing the categorisation of classes.

12.4.2 Feature Extraction, ML/DL Model Selection and Training

Both NEAI and EI offer an automated mechanism for generating the AI model architecture and training, although the mechanisms differ since NEAI employs unsupervised algorithms, whereas EI employs DL NNs.

The benchmarking process for n-class classification with NEAI involves searching through a pool of ML algorithms and tests combinations of three elements: pre-processing, ML algorithms (e.g., random forest-RF, support vector machines-SVM, etc.) and hyper-parameters for each model. Each combination results in a library that is evaluated for accuracy, confidence and memory usage, and the results provide a ranking of these libraries. Accuracy reflects the library's ability to correctly attribute each signal to the correct class, whereas confidence reflects the library's ability to separate the n-classes.

Figure 12.4 shows that the top library for the PdM classification case has an accuracy of 100%, confidence 99.94%, uses the RF algorithm, and takes 6.2kB RAM and 8.3 kB Flash. 100% means that all classes are completely separated, there is no overlap.

In the "Confusion Matrix", the 200 number means that the performance for each class is 100%, i.e., all 200 signals extracted from initial data (20% of 1000 signals) have been properly classified.

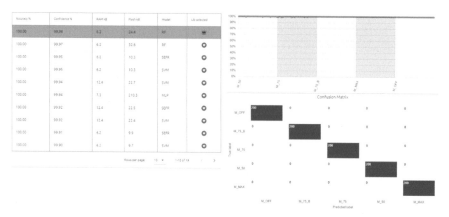

Figure 12.4   Benchmarking with NEAI

In the EI platform, a *Spectral Analysis* signal processing block was used to apply a filter, perform spectral analysis, and extract frequency and spectral power data. A useful aspect of the platform is the possibility to visualise and explore the features (Figure 12.5). The fact that the features are visually clustered is a good indication that the model can be trained to perform the classification. During the first iterations, the features overlapped to a significant degree and were intertwined, and the trained model had difficulties differentiating between classes. This problem was addressed by collecting more signals and increasing the size of the sampling signal to better capture signal patterns.

It is also possible to calculate and visualise feature importance when generating the features, indicating how important the features are for each class compared with all other classes. RMS and peak values of vibration along the three-axis proved to be the most important features in determining the class in this case. Based on this information, the dimension reduction algorithms can be used to simplify the model by deleting the less important or redundant information from the data set to make it manageable while maintaining relevance and performance.

To implement the solution in EI, a classification learning block was used, which employs TensorFlow with Keras. It takes the features from *Spectral Analysis* signal processing block and learns to distinguish between the five classes. The strategy adopted was to start with a small deep NN model and experiment with it, i.e., two dense layers, using EI graphical user interface (GUI). Most of the experimentations have been performed around an architecture consisting of multiple dense layers and dropout layers. Convolutional layers were also included.

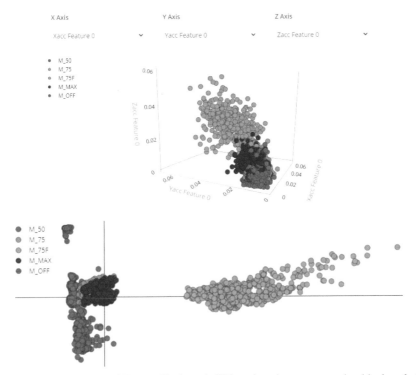

Figure 12.5   Snapshots of Feature Explorer in EI based on the pre-processing block early in the process.

At the end of the training, the model's performance and the confusion matrix of the validation data can be evaluated. Figure 12.6 shows an accuracy and a loss on the training and validation datasets, comparable with the results obtained with NEAI with a different model architecture. To avoid overfitting, the learning rate was reduced, and more data was collected, and the model was re-trained.

12.4.3 Optimisation and Tuning Performance

Developing the most efficient ML/DL flows for the classification PdM application was challenging. It required many iterative experiments and insights into the workings of motor vibration patterns, digital signal processing, AI algorithms, architectures, and microcontrollers. Nevertheless, both NEAI and EI provided automation and transparency for these processes, though to varying degrees.

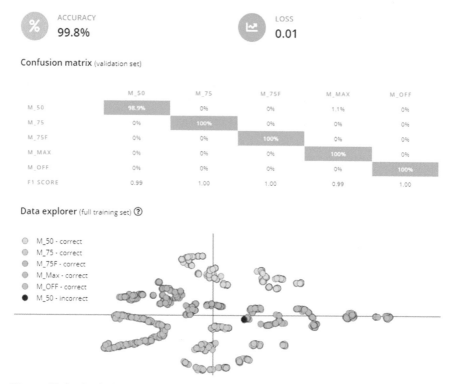

Figure 12.6 Confusion Matrix and Data Explorer based on full training set: Correctly Classified (Green) and Misclassified (Red).

For the NEAI classification, the learning is fixed at library generation based on the data provided for each class. The benchmarking implementation includes patented elements; thus, the internal working of the engine is not transparent. Nevertheless, multiple benchmarks can be created, and a high degree of automation allows for the best results to be obtained from signal capturing and formatting. The benchmarking process takes around 60 minutes when running on a processing unit with 6 CPU cores.

EI offers a higher degree of transparency and control over the model architecture and hyperparameters. The strategy adopted for the case of EI was to start from a simple model, experiment with it and improve it into a deeper and wider model. For this improvement step and for validation purposes, a parallel sub-flow was branched out from the flow with EI to conduct experiments in a Python framework. The training was launched in both EI and Python and compared throughout. The updated architecture and

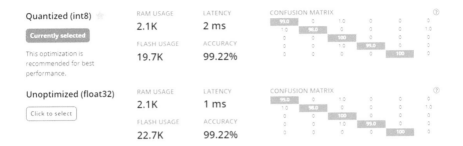

Figure 12.7 A comparison between int8 quantized and unoptimized versions of the same model, showing the difference in performance and results.

hyperparameters were exchanged back and forth between the EI and Python frameworks.

The improvements consisted in making the model deeper by adding more layers, and wider by increasing the number of hidden units, changing the activation and optimisation functions, learning rate, fitting more data.

While the improvement process was run manually in Python, the EI's Edge Optimized Neural (EON^{TM}) Compiler [9] can be used to find the best solution for the Arm® Cortex®-M-based MCUs, i.e., the most optimal combination of processing block and ML model for the given set of constraints, including latency, RAM usage, and accuracy. Currently, there are a limited number of MCUs that are supported and does not include the MCU of STWIN IIoT device (Arm® Cortex®-M4 MCU STM32L4R9), which operates at a frequency of up to 120MHz. Nevertheless, the estimated on-device performance could be evaluated for Cortex-M4F 80MHz, to determine the impact of optimisations such as quantisation across different slices of the datasets (Figure 12.7).

12.4.4 Testing

ML/DL model testing usually refers to the evaluation of the trained model on the testing dataset to analyse how well the model performs against unseen data. However, model testing in NEAI and EI provide more than that. Both platforms provide a microcontroller emulator to test and debug the generated model prior to its deployment on the device.

As part of the NEAI toolkit, a microcontroller emulator is provided for each library to test and debug the generated model without the need to download, link or compile. Test signals can be imported from file; however,

Figure 12.8   Evaluation of trained model using NEAI Emulator with live streaming.

the signals were imported live from the same datalogger application through serial port, in this way ensuring completely new signals, not seen before. The classification is automatically run using the live signals, while changing motor speeds and triggering shaft disturbances, to switch between classes and cover all five states and classes.

The results are presented in Figure 12.8, showing that the classifier managed to properly reproduce and detect all classes with reasonable certainty percentages.

In EI, the trained model was evaluated by assessing the accuracy using the test dataset. To ensure unbiased evaluation of model effectiveness, the test samples were not used directly or indirectly during training. The EI emulator took care of extracting the features from the test set, running the trained model, and reporting the performance in the confusion matrix. The results are shown in Figure 12.9.

12.4.5 Deployment

In the context of micro-edge embedded systems, model deployment is dependent on the hardware/software platform and is more or less automated, and in essence comprises three steps: the first is a format conversion of the fully trained model; the second is a weight/model compression to reduce the amount of memory to store the weights in the target hardware platform and to simplify the computation so it can run efficiently on target processors. The third entails compiling the model and generating the code to be integrated with the MCUs firmware.

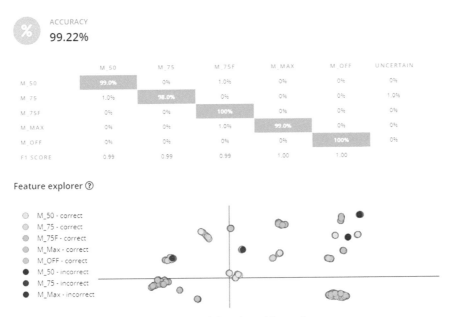

Figure 12.9 EI model testing with test datasets.

The back-end flow consists of wrapping an STM32CubeIDE project with the generated files from the three deployed models, adding functionality on top such as retrieving the accelerometer values to be fed to the classification function and displaying the result, then compiled, built, and flashed onto the MCU target.

The flow exhibits some particularities in the case of the three model deployments.

In the case of NEAI, the selected model is deployed in the form of a static library (libneai.a), an AI header file (NanoEdgeAI.h) containing functions and variable definitions, and a knowledge header file (knowledge.h) containing the model's knowledge. In this case, first the knowledge was initialised, then the NanoEdge AI classifier was run, and the output was print to the serial port.

For the EI deployment, the CMSIS-PACK [11][12] for STM32 packaged all signal processing blocks, configuration and learning blocks up into a single library (.pack file), which was then added to the STM32 project using the CubeMX packages manager. This is currently only supported for C++ applications using CubeIDE.

Detected class : MOTOR_OFF Certainty: 99%
Detected class : MOTOR_OFF Certainty: 99%
Detected class : MOTOR_OFF Certainty: 99%
?
Detected class : MOTOR_OFF Certainty: 99%
Detected class : MOTOR_ON_NORMAL_50 Certainty: 91%
?
Detected class : MOTOR_OFF Certainty: 99%
Detected class : MOTOR_OFF Certainty: 99%
?
Detected class : MOTOR_ON_NORMAL_75 Certainty: 91%
Detected class : MOTOR_ON_NORMAL_75 Certainty: 86%
Detected class : MOTOR_ON_NORMAL_75 Certainty: 63%
Detected class : MOTOR_ON_NORMAL_MAX Certainty: 76%
Detected class : MOTOR_ON_NORMAL_75 Certainty: 95%
?
Detected class : MOTOR_ON_NORMAL_75_B Certainty: 98%
Detected class : MOTOR_ON_NORMAL_75_B Certainty: 97%
Detected class : MOTOR_ON_NORMAL_75_B Certainty: 97%

Figure 12.10 Live classification streaming with detected state and confidence (with Tera Term)

The third flow was branched out from EI and further developed in a Python framework using TensorFlow's Keras API. The resulted model was converted into optimised C code with STM32 Cube.AI, an extension of the CubeMX tool, which offers simple and efficient interoperability with other ML frameworks.

12.4.6 Inference

Inference classifications have been conducted with all applications running directly from the target hardware platform on the micro-edge IIoT devices, producing classification in real-time.

The state machine consists mainly of two states with two functions "init" and "inferencing", respectively, with the former initialising the deep NN model and the latter being a continuously running function for collecting raw data from the sensors on the micro-edge IIoT device and making classifications in real-time. A snapshot from the classification based on the NEAI model is shown in Figure 12.10.

The "?" indicate the state switching, which happens after several consecutive confirmations of inference result is encounter, and this number is programmable.

12.5 Discussion and Future Work

Embedding trained models into the firmware code enables AI/ML capabilities of intelligent edge devices. Employing different frameworks that permit the integration of complex AI mechanisms within MCUs - such as NEAI Studio, EI and STM32 Cube.AI - for deploying AI-based PdM solutions into micro-edge embedded devices provides designers with the flexibility to optimise implementation by experimenting with deployment on the same hardware platform target using several frameworks and inference engines. The different workflows can be matched to the PdM application requirements for generating embedded code and performing learning and inference engine optimisations.

ML and NNs can now be efficiently deployed on resource-constrained devices, which enable cost-efficient deployment, widespread availability, and the preservation of sensitive data in PdM applications. However, the trade-offs associated with optimisation methods, software frameworks and hardware architecture on performance metrics, such as inference latency and energy consumption, are yet to be studied and researched in depth.

This preliminary work allowed for the exploration of different scenarios to evaluate trade-offs between computational cost and performance on actual classification tasks, laying the foundation for further investigations of more complex PdM systems using various AI-based techniques. Future work will aim to enlarge comparison and benchmarking by considering more edge ML and DL technologies, workflows, and datasets. A more generic and complete PdM strategy must include insights from other applications, such as anomaly detection, regression, and forecasting.

Acknowledgements

This work is conducted under the framework of the ECSEL AI4DI "Artificial Intelligence for Digitising Industry" project. The project has received funding from the ECSEL Joint Undertaking (JU) under grant agreement No 826060. The JU receives support from the European Union's Horizon 2020 research and innovation programme and Germany, Austria, Czech Republic, Italy, Latvia, Belgium, Lithuania, France, Greece, Finland, Norway.

References

[1] R. Sanchez-Iborra and A.F. Skarmeta, "TinyML-Enabled Frugal Smart Objects: Challenges and Opportunities," in *IEEE Circuits and Systems Magazine*, vol. 20, no. 3, pp. 4-18, third quarter 2020. https://doi.org/10.1109/MCAS.2020.3005467

[2] T. Hafeez, L. Xu and G. Mcardle, "Edge Intelligence for Data Handling and Predictive Maintenance in IIoT," in *IEEE Access*, Vol. 9, pp. 49355-49371, 2021. https://doi.org/10.1109/ACCESS.2021.3069137

[3] Y. Liu, W. Yu, T. Dillon, W. Rahayu and M. Li, "Empowering IoT Predictive Maintenance Solutions With AI: A Distributed System for Manufacturing Plant-Wide Monitoring," in *IEEE Transactions on Industrial Informatics*, vol. 18, no. 2, pp. 1345-1354, Feb. 2022. https://doi.org/10.1109/TII.2021.3091774

[4] H. Wang, H. Sayadi, S.M. Pudukotai Dinakarrao, A. Sasan, S. Rafatirad and H. Homayoun, "Enabling Micro AI for Securing Edge Devices at Hardware Level," in *IEEE Journal on Emerging and Selected Topics in Circuits and Systems*, vol. 11, no. 4, pp. 803-815, Dec. 2021. https://doi.org/10.1109/JETCAS.2021.3126816

[5] F. Cipollini, L. Oneto, A. Coraddu, et al. "Unsupervised Deep Learning for Induction Motor Bearings Monitoring". Data-Enabled Discov. Appl. 3, 1, 2019. https://doi.org/10.1007/s41688-018-0025-2

[6] M. Guenther. 6 Ways to Improve Electric Motor Lubrication for Better Bearing Reliability. Available online at: https://blog.chesterton.com/lubrication-maintenance/improving-electric-motor-lubricaiton/

[7] C. Kammerer, M. Gaust, M. Küstner, P. Starke, R. Radtke, and A. Jesser, "Motor Classification with Machine Learning Methods for Predictive Maintenance," *IFAC-PapersOnLine*, vol. 54, no. 1, pp. 1059–1064, 2021. https://doi.org/10.1016/j.ifacol.2021.08.126

[8] Edge Impulse. Available online at: https://www.edgeimpulse.com

[9] EON Tuner. Available online at: https://docs.edgeimpulse.com/docs/eon-tuner

[10] J. Jongboom, 2020. "Learning for all STM32 developers with STM32Cube.AI and Edge Impulse". Available online at: https://www.edgeimpulse.com/blog/machine-learning-for-all-stm32-developers-with-stm32cube-ai-and-edge-impulse

[11] ARM-NN. 2020. Available online at: https://github.com/ARM-softwar e/armnn

[12] CMSIS-NN. 2020. Available online at: https://arm-software.github.io/C MSIS_5/NN/html/

[13] STM32Cube.AI 2020. Available online at: https://www.st.com/en/embe dded-software/x-cube-ai.html

[14] NanoEdge^{TM} AI Studio. Automated Machine Learning (ML) tool for STM32 developers. Available online at: https://www.st.com/en/develo pment-tools/nanoedgeaistudio.html

13

AI-Driven Strategies to Implement a Grapevine Downy Mildew Warning System

Luiz Angelo Steffenel[1], Axel Langlet[1], Lilian Hollard[1], Lucas Mohimont[1], Nathalie Gaveau[1], Marcello Copola[2], Clément Pierlot[3], and Marine Rondeau[3]

[1]Université de Reims Champagne Ardenne, France
[2]STMicroelectronics, France
[3]Vranken-Pommery Monopole, France

Abstract

In this paper, we assess the usage of machine learning techniques to predict the infection events of Downy Mildew. Every year, Champagne vineyards are exposed to grapevine diseases that affect the plants and fruits, most caused by fungi. Using data from an agro-meteorological station, we compare machine learning performances against traditional prediction methods for Downy Mildew (*Plasmopara viticola*) infections. Indeed, depending on the year, we obtain 82 to 97% accuracy for primary infections and 98% for secondary infections. These results may guide the development of Edge AI applications integrated to meteorological stations and agricultural sensors,and help winegrowers to rationalize the vine's treatment, limiting the damages and the usage of fungicide or chemical products.

Keywords: artificial intelligence, Downy Mildew, CNN, random forest, SVM.

13.1 Introduction

Every year, Champagne vineyards are exposed to grapevine fungal diseases that affect the plants and fruits. Black rot (*Guignardia bidwellii*), Downy

177

DOI: 10.1201/9781003377382-13

mildew (*Plasmopara viticola*), Powdery mildew (*Erysiphe necator*), and Graymold (*Botrytis cinerea*) are examples of diseases that can affect grape quality and hinder the productivity. Each fungus develops under certain environmental conditions and detecting favourable conditions for the spread of the diseases may lead to proactive actions to prevent its dissemination.

In the specific case of the Downy Mildew caused by *Plasmopara viticola*, there are two cycles of infestation that affect the grapevine. The first one is caused by sexual spores (called *primary infections*) and the second one by the dissemination of asexual (*secondary infections*) [4].

The mechanical identification of the fungus development cycles and their forecast has already been the subject of several works, including [8][5] or [7]. Indeed, several of these works define algorithms to identify the primary or secondary infection events using a combination of weather and ground observed variables, which led to the creation of decision-support systems for the vine-growers. However, these algorithms are limited to strict input parameters, which are not always available, and do not explore the potential of hidden correlations with other data variables such as dew point, cloud coverage or vapor pressure deficit.

Artificial intelligence, on the other side,relies only on the dataset rather than on models. It uses computing power to expand the search for patterns and correlations among a broader and richer dataset, often reaching similar or better results than existing models.

Despite its potential, artificial intelligence has been rarely used to identify Downy Mildew infections. Among the precursor works, we can cite Chen et al. [3], which applied several regression models as well as random forest and gradient boost to predict severe infection events in the Bordeaux vine-yard. Volpi et al. [9] also use decision trees and random forests to identify different diseases in Tuscany, Italy, but relying on meteorological data from ERA5-Land instead of in-site sensors.

Interestingly, artificial intelligence is more used to monitor crops through image systems rather than weather sensors. For instance, [1][2] use image recognition techniques to identify the intensity of the infections on water-melon or squash crops using hyperspectral images from aerial views. Another work [6] uses Convolutional Neural Networks to detect *Plasmopara viticola spores* in microscopic images.

In this paper, we explore the interest of using machine learning techniques to identify Downy Mildew infections using datasets obtained from regular agro-meteorological sensors. Our aim is both to identify the most efficient and robust methods and to prepare the path to their implementation on Edge AI devices deployed directly on the vineyards.

The remainder of this paper is organized as follows: Section 13.2 presents the datasets and research methodology used in this work. Section 13.3 introduces the different machine learning techniques used in this work, as well as their implementation specifications. In Section 13.4 we present a comparative study of machine learning strategies, aiming at their accuracy as well as their robustness over the years. Section 13.5 goes beyond the simple results by discussing the impact of AI-based algorithms on the monitoring of crops. Finally, Section 13.6 concludes this work.

13.2 Research Material and Methodology

13.2.1 Datasets

The data used in this paper was obtained from a Promété AGRI-300 weather station installed at "Moulin de la Housse" vineyard from Vranken-Pommery group in Reims[1]. This station provides hourly readings from several features of interest:

- Wind speed [Km/h] (max, average)
- Wind gust [Km/h] (max)
- Relative humidity [%] (max, min, average)
- Pluviometry [l/m^2]
- Leaf wetting duration [min]
- Dew point [C] (min, average)
- Solar radiation [W/m^2] (average)
- Air temperature [C] (max, min, average)
- Vapor press deficit [kPa] (min, average)

More than 20k entries were recorded for each feature from 2019 to 2021, except for the Leaf wetting duration that could only be recorded in 2019/2020 as the sensor stop working in February 2021.

The presented machine learning approaches are implemented, optimized and evaluated on a Nvidia DGX1 server that includes eight Tesla V100 GPUs connected through an NVlink network supporting up to 40 GB/s bidirectional bandwidth. Regarding programming tools, we have implemented our approaches using the Python language with scikit-learn, Tensorflow and Keras libraries.

---

[1]Data could be provided upon request

13.2.2 Labelling Methodology

To train machine learning models to identify Mildew favourable situations, we adopted a supervised learning approach. To label the training dataset, we applied the algorithms proposed by [7]. Two different Mildew infection alert situations are identified in that work, each one with strict requirements. Hence, primary infections are related to the conditions for winter spores' germination, which may occur when the average daily temperature exceeds 10 °C and the precipitation within the last 48h reaches 10 mm (called "3-10" flag). If rainfall or gentle breeze (i.e., wind of speed greater than 3.4m/s) occurs at night within the following 48h, primary infection has presumably occurred, causing the start of the incubation period of *Plasmopora viticola*. Figure 13.1 schematizes this algorithm.

Second mildew infections may happen when the incubation period from the first infection has been completed. It depends on favourable night conditions (FNCs) conditions where the weather is humid (relative humidity (RH) >80%), and the temperature is higher than 12°C for at least 2h. In such case, the secondary infection warning is raised if we also observe more than 2h of uninterrupted leaf wetness (LW) and average temperature (T) above 10°C, with precipitation or strong wind that can increase spore spread. Figure 13.2 schematizes this algorithm.

Thanks to these two algorithms, we create two binary labels, one for primary alert and the other for secondary alert, used in independent classification models. These labels are only used during the training phase, as our objective is to obtain accurate predictions based on the raw input data from the weather station sensors.

13.3 Machine Learning Models

This section presents different strategies to model the Downy Mildew warning system using machine-learning techniques. As presented in Section 13.2, our dataset covers three years (2019-2021) and includes several features directly related to the algorithms from [7] such as temperature, relative humidity, pluviometry, wind speed or leaf wetness. Other algorithms variables were adapted from existing data, so the absence of solar radiation (provided by the weather station) was used as an indicator for night time instead of a calculation based solely on the date.

We deliberately kept other variables not cited in the original algorithms, such as the dew point and the vapor press deficit. As stated before, our aim

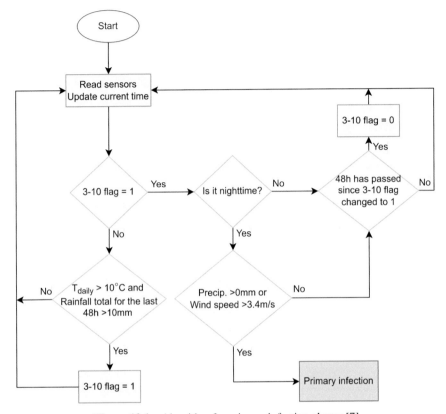

Figure 13.1 Algorithm for primary infection alarms [7]

is to explore potential correlations with additional variables. Similarly, we do not compare the accuracy with the real risks in the vineyard but only with the expected labels. Performing such comparison requires on-site evaluation and a separate tagging from a human operator, which is part of our future works.

Another point to consider is how to enter the dataset as alerts depend on historical events from at least the last 48h. Instead of using mode complex time-series models such as LSTM or GRU, we chose to feed the algorithms with a concatenation of the features recorded in the last 48h. This approach allows us to express the problem in a simpler way that can be approached using a wider range of machine learning techniques, including some best adapted to constrained environments such as those in a Edge AI scenario.

As a result, we model the problem as a binary classification problem, i.e., for each level of infection alert (primary or secondary), we create separated "

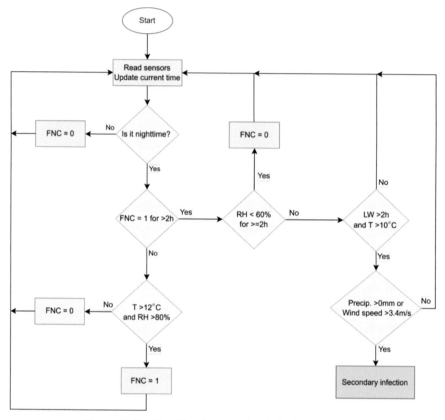

Figure 13.2 Algorithm for secondary infection alarms [7]

alert"/ "not alert" labels. We decided to split it into two binary classification problems instead of a multi-class classification problem to favour each alert type's accuracy. Henceforth, we choose to compare five well-known binary classification techniques:

- Decision trees
- Random forest
- Support Vector Machines (SVM)
- Dense Neural Networks (DNN)
- Convolutional Neural Networks (CNN)

Decision Trees and Support Vector Machine predictors use the basic *scikit-learn* implementation (*DecisionTreeClassifier* and *SVC*, respectively) with no additional optimisation. Random Forests (*RandomForestClassifier*) were trained with the parameter *n iterators = 1000*.

The Dense Neural Network implemented in Keras using seven Dense layers with respectively 200, 100, 100, 50, 50,10, and 2 outputs. Activator *ReLU* was used in all but the last layer (*None*); the model was compiled with *Binary Crossentropy (from logits=True)* loss, Adam optimizer, and *Binary Accuracy* metrics.

Finally, the Convolutional Neural Network was implemented in Keras, using at the input two conv2D layers (32 and 64 outputs, respectively), with 3x3 padding, 0.2 dropout and ReLU activation. Once flatted, a Dense network with 100 outputs, 0.5 dropout and ReLU activation sits just before a final Dense network with 2 outputs (Sigmoid activation).

As the dataset only covers three years, we adopted a cross-validation approach where, for each technique, we generated a different model for each respective year (2019, 2020 or 2021). Therefore, each model was trained only with the data from its own year, split into 90% training and 10% testing parts (randomly shuffled) and later submitted to cross-validation against the other years. Not only the cross-validation helps identifying the most robust model but also allows to investigate the impact of the 2021 weather profile, which differed from the two previous ones due to several climatic events (early crop freeze, rainy weather) that favoured the spread of diseases and led to a massive reduction in crop production and quality.

13.4 Results

13.4.1 Primary Mildew Infection Alerts

As stated above, we create three different training-validation datasets, one for each year. Therefore, Table 13.1 compares the accuracy score from the 2019's model when applied to 2020 and 2021. The best scores are presented in bold, showing that two techniques detach from the others: CNN and SVM. CNN shows slight better scores in the 2021 dataset but is closely followed by SVM.

In the case of the 2020's model, Random Forest and SVM perform well for the 2019 case, and almost all techniques (except simple Decision Tree) present similar results for the 2021 case (see Table 13.2). Finally, the 2021's model Random Forest seems the best technique for the 2019 dataset, while SVM is better in the case of the 2020 dataset (Table 13.3). We can, however, point out that Random Forest achieves good results in this latter case, even if not as good as the SVM scores. If the" best" technique varies from year to

Table 13.1   Accuracy of 2019 Primary Infection Models

	2019	2020	2021
Decision Tree	-	0.607	0.597
Random Forest	-	0.841	0.743
Support Vector Machines	-	**0.978**	0.821
Dense Neural Network	-	0.909	0.815
Convolutional Neural Network	-	**0.978**	**0.822**

Table 13.2   Accuracy of 2020 Primary Infection Models

	2019	2020	2021
Decision Tree	0.925	-	0.797
Random Forest	**0.935**	-	**0.821**
Support Vector Machines	**0.935**	-	**0.821**
Dense Neural Network	0.925	-	**0.821**
Convolutional Neural Network	0.932	-	**0.821**

Table 13.3   Accuracy of 2021 Primary Infection Models

	2019	2020	2021
Decision Tree	0.921	0.881	-
Random Forest	**0.938**	0.961	-
Support Vector Machines	0.935	**0.974**	-
Dense Neural Network	0.910	0.849	-
Convolutional Neural Network	0.915	0.895	-

year, both SVM and CNNs show robust results, closely followed by Random Forest. The choice reposes therefore in the computing capabilities available to the devices.

We can also see that 2021 was different from the previous ones. If models from 2019 or 2020 achieve lower scores when predicting 2021 alerts, we can also say that models trained with 2021 data are among the best ones when predicting alerts for the previous years. This was somehow expected, as 2021 was rich in favourable events for spreading diseases in the vineyard.

13.4.2 Secondary Mildew Infection Alerts

As the meteorological station stopped recording leaf wetness in February 2021, we could not tag Secondary Mildew Infections on the 2021 dataset. Nonetheless, we compare both 2019 and 2021 models in cross-validation, as we previously did for the Primary Mildew Infection.

Hence, Table 13.4 condenses the results from all machine learning techniques when cross validating each year's models. Secondary Infection alerts

Table 13.4 Accuracy of 2021 Primary Infection Models

	2019 (model 2020)	2020 (model 2019)
Decision Tree	0.960	0.895
Random Forest	0.979	0.988
Support Vector Machines	0.979	**0.991**
Dense Neural Network	0.932	**0.991**
Convolutional Neural Network	**0.980**	**0.991**

seem much easier to identify, with higher accuracy scores. Unfortunately, the absence of a 2021 dataset does not allow a broader comparison under different weather conditions (2021 presented the lowest accuracy in the Primary Alert experiments).

Once again, CNN presents the highest accuracy scores, closely followed by SVM and Random Forest. Indeed, we shall point-out that SVN and Random Forest are good candidates when considering the implementation on environments with performance restrictions, such as in the case of IoT / Edge AI.

13.5 Discussion

The results obtained here are encouraging but shall be considered in the context of the reduced span of the dataset gathered from a single agro-meteorological station installed since 2019. A deeper analysis would require several years of data, as performed by [3] or [9].

However, our main objective was to conceive a proof of concept inscribed in the efforts of the European project AI4DI to develop and disseminate an environmental monitoring system based on different industrial sensors (e.g., TEROS, Bosch BME68x, ST Microelectronics) connected to STM32WL enhanced by a machine learning core. These sensors are expected to enable continuous monitoring of the environment, the soil, meteorological conditions, and/or plant performances.

Besides implementing AI models on the STM32WL, some sensors can also be enriched with a machine learning core. This is the case of the LSM6D SOX sensor from ST Microelectronics, which comprises a set of configurable parameters and decision trees able to run AI algorithms in the sensor itself. Hence, this environment would benefit from simpler models such as random forest and SVM, rather than CNN.

Today, while many agricultural weather meteorological stations are available on the market, innovation comes from implementing Edge AI directly on the sensors or, in some cases, in the gateways. Therefore, the current work represents a primary effort to identify good and robust models that could be deployed in an edge AI environment.

13.6 Conclusion

Every year, Champagne vineyards are exposed to grapevine diseases that affect the plants and fruits, and the Downy Mildew, caused by *Plasmopara viticola* is a common disease. Forecasting the infection events of Downy Mildew may help vine growers to rationalize the treatment of the vine, limiting the damages and the usage of fungicide or chemical products.

In this paper, we compare the accuracy of several machine learning techniques when applied to datasets from the Champagne region. By creating multiple models and using cross-validation across different years, we were able to identify three candidate techniques with close results, namely Convolutional Neural Networks, Support Vector Machines and Random Forest.

If CNN seems to be more robust across different years, the accuracy difference is minimal,and the other techniques present an interest in the case of deployment over an Edge AI infrastructure. Indeed, we aim to prepare the path to the implementation of Downy Mildew forecast models on Edge AI sensing devices that will be deployed directly on the vineyards to closely monitor the crops.

Acknowledgements

This work has been performed in the project AI4DI: Artificial Intelligence for Digitizing Industry, under grant agreement No 826060. The project is cofunded by grants from Germany, Austria, Finland, France, Norway, Latvia, Belgium, Italy, Switzerland, and the Czech Republic and - Electronic Component Systems for European Leadership Joint Undertaking (ECSEL JU).

We want to thank Vranken-Pommery Monopole for providing the datasets for the training. We also thank the ROMEO Computing Center[2] of Université de Reims Champagne Ardenne, whose Nvidia DGX-1 server allowed us to accelerate the training steps and compare several model approaches.

---

[2]https://romeo.univ-reims.fr

References

[1] J. Abdulridha, Y. Ampatzidis, J. Qureshi, and P. Roberts. Identification and classification of downy mildew severity stages in watermelon utilizing aerial and ground remote sensing and machine learning. *Frontiers in Plant Science*, 13, 2022.

[2] J. Abdulridha, Y. Ampatzidis, P. Roberts, S. C. Kakarla. Detecting powdery mildew disease in squash at different stages using UAV-based hyperspectral imaging and artificial intelligence. *Biosystems Engineering*, 197:135–148, 2020.

[3] M. Chen, F. Brun, M. Raynal, and D. Makowski. Forecasting severe grape downy mildew attacks using machine learning. *PLOS ONE*, 15:1–20, 03 2020.

[4] C. Gessler, I. Pertot, and M. Perazzolli. Plasmopara viticola: A review of knowledge on downy mildew of grapevine and effective disease management. *PhytopathologiaMediterranea*, 50:3–44, 04 2011.

[5] E. Gonzalez-Domínguez, T. Caffi, N. Ciliberti, and V. Rossi. A mechanistic model of botrytis cinerea on grapevines that includes weather, vine growth stage, and the main infection pathways. *PLOS ONE*, 10(10):1–23, 10 2015.

[6] I. Hernández, S. Gutiérrez, S. Ceballos, R. Iñiguez, I. Barrio, and J. Tardaguila. Artificial intelligence and novel sensing technologies for assessing downy mildew in grapevine. *Horticulturae*, 7(5), 2021.

[7] I. Mezei, M. Lukic, L. Berbakov, B. Pavkovic, and B. Radovanovic. Grapevine downy mildew warning system based on nb-iot and energy harvesting technology. *Electronics*, 11(3), 2022.

[8] V. Rossi, T. Caffi, S. Giosue, and R. Bugiani. A mechanistic model' simulating primary infections of downy mildew in grapevine. *Ecological Modelling*, 212(3):480–491, 2008.

[9] I. Volpi, D. Guidotti, M. Mammini, and S. Marchi. Predicting symptoms of downy mildew, powdery mildew, and graymold diseases of grapevine through machine learning. *Italian Journal of Agrometeorology*, (2):57–69, Dec. 2021.

14

On the Verification of Diagnosis Models

Franz Wotawa and Oliver Tazl

Graz University of Technology, Austria

Abstract

Enhancing systems with advanced diagnostic capabilities for detecting, locating, and compensating faults during operation increases autonomy and reliability. To assure that the diagnosis-enhanced system really has improved reliability, we need – besides other means – to check the correctness of the diagnosis functionality. In this paper, we contribute to this challenge and discuss the application of testing to the case of model-based diagnosis, where we focus on testing the system models used for fault detection and localization. We present a simple use case and provide a step-by-step discussion on introducing testing, its capabilities, and arising issues. We come up with several challenges that we should tackle in future research.

Keywords: model-based diagnosis, testing, verification and validation.

14.1 Introduction

Every system comprising hardware faces the problem of degradation under operation, which impacts its behavior over time. To prevent unwanted behavior that may lead to harm, we have to carry out regular maintenance tasks. Maintenance includes preventive activities like changing the tires of cars when their surfaces do not meet regulations anymore and looking at errors occurring during operation. The latter requires root cause identification, i.e., searching for components we have to repair for failure recovery. There is no doubt that the maintenance and diagnosis of engineered systems are of practical importance and, therefore, worth being considered in research.

DOI: 10.1201/9781003377382-14

If we aim to support maintenance personnel carrying out diagnoses, we need to automate the fault detection and localization activities. Since the beginning of artificial intelligence, diagnosis has been an active research field leading to expert systems and later to model-based diagnosis. The idea behind model-based diagnosis is to use system models for localizing the root causes of detected failures. Early work includes Davis and colleagues [3] papers discussing the basic ideas and concepts behind model-based reasoning. Later, Reiter [15] formalized the idea utilizing first-order logic. Based on these foundations, several authors have discussed several applications of model-based reasoning for solving real-world problems. Applications range from power supply networks [1], the automotive domain [13], space probes [14], robotics [7], self-adaptive systems [16], to debugging [6]. For a more recent paper, we refer to Wotawa and Kaufmann [22], where the authors introduced how advanced reasoning systems can be used for computing diagnosis. For recent applications of diagnosis in the context of cyber-physical systems, have a look at [9, 23, 21, 20].

In the following, we illustrate the basic ideas and concepts of model-based reasoning using a small example circuit comprising a battery B, a switch S, and two bulbs L_1, L_2. We depict the circuit in Figure 14.1. If we switch on S, we expect both bulbs to transmit light when we assume the correctness of every component. It is important to consider such correctness assumptions. For example, if we switch on S, and only one bulb (e.g., L_1) is on, and the other (e.g., L_2) is not, we conclude a broken bulb. But how can we do this? We may consider a model for each component, e.g., a correct battery provides electricity, a switch in the on state takes the electricity from the battery and transmits it to the bulbs, and a correct bulb produces light if there is electricity available. When we assume that all components are working, we receive a contradiction from this model. This is due to bulb L_2 that should produce light but we do not observe it. If we assume all components except L_2 to be correct, there is no contradiction anymore, and we have identified the root cause, i.e., L_2.

A prerequisite of model-based diagnosis is the availability of a system model (or model in short). Modeling is not a trivial task. For model-based diagnosis, we need models formulated in a language that a reasoning system can use for deriving logical conclusions. Models are abstract representations of the system structure and behavior. Only parts of the system classified as components in the model can be part of a derived root cause. Wires or connectors need to be stated as components if we want to have them included in a diagnosis. In model-based diagnosis, only components used in models

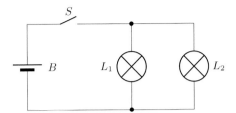

Figure 14.1   A simple electric circuit comprising bulbs, a switch and a battery.

can be part of a root cause. It is also worth noting that we can use uncertainty in model-based diagnosis. De Kleer and Williams [4] formalized the use of fault probabilities of components for searching for the most probable diagnosis. In addition, de Kleer and Williams introduced an algorithm for selecting the optimal probing locations for minimizing probing steps for identifying a single diagnosis.

In this manuscript, we do not focus on the diagnosis methods and processes themselves. Instead, we provide a discussion on how to verify diagnosis models. The challenge of model verification is of uttermost importance for assuring that systems equipped with diagnosis functionality work correctly. Although we may use some of the presented results for verifying diagnosis models generated by machine learning, we consider models for model-based reasoning in the context of this paper. For testing machine learning, we refer the interested reader to a recent survey [24].

The challenge of model-based diagnosis and other logic-based reasoning systems is not that novel. Wotawa [17] introduced the use of combinatorial testing and fault injection for testing self-adaptive systems based on models. The same author also discussed the use of combinatorial testing and metamorphic testing for theorem provers in [18] and the general challenge [19]. In any of these papers, the focus is on testing the implementation and not the underlying models. Koroglu and Wotawa [10] also contributed to the challenge of verifying the reasoning system but focused on the underlying compiler that allows reading in logic theories, i.e., system models. Hence, testing the system models used for diagnosis is still an open challenge worth tackling for quality assurance.

We organize this paper as follows: In Section 14.2, we introduce the testing challenge in detail including a first solution. Afterward, we present the results when using the provided solution in a small case study. Finally, we discuss open issues, and further challenges, and conclude the paper.

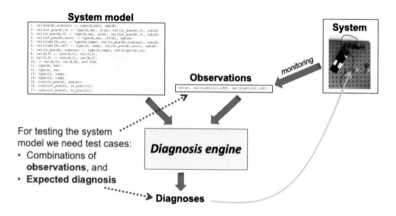

Figure 14.2 The model-based diagnosis principle and information needed for testing.

14.2 The Model Testing Challenge

Before discussing the model testing challenge in detail, we briefly summarize model-based diagnosis and the required information. In Figure 14.2 we depict the basic architecture behind every model-based diagnosis system. On the right side, we have a (physical) system from which we extract observations. On the upper left side, we have a model of the system. This model shall represent the system in a way such that expected observations can be derived. The model and the observations are passed to a diagnosis engine, which tries to find an arrangement of health states to components such that no contradiction can be derived. In the simplest case, we only know the correct behavior of components. We use a logic predicate nab\1 to represent the corresponding health state. The diagnosis engine itself is assumed to be based on either a theorem prover or a constraint solver. It delivers a set of diagnoses. Each diagnosis itself is a set of faulty components. If the set of diagnoses comprises the empty set, we know that all components are working as expected.

It is worth noting that in the context of this paper, we are not interested in outlining the details regarding model-based diagnosis, the modeling principles, and algorithms. We solely focus on testing, and specifically on testing the system model. What we can take with us from Figure 14.2 are the inputs and outputs to the diagnosis engine comprising the model, the observations, and the computed diagnoses. If we want to verify the implementation of the diagnosis engine, we can use models and observations together with the corresponding expected diagnoses to define a test case. However, when we want to test the models, which are usually divided into two parts, the component models, and the structure of the system, we have to further think about underlying assumptions and prerequisites.

First, we have to assume that the diagnosis engine itself is correct. This means that the diagnosis engine is delivering the right diagnoses for a given model and observations. Testing the implementation of the diagnosis engine might also comprise testing the underlying theorem prover or constraint solver, the implementation of the diagnosis algorithm, and the compiler that is used to load a model and the observations into the diagnosis engine.

Second, the observations themselves describe the data that have been observed from the system. Usually, we do not use the raw data obtained from the system directly. The data is usually mapped to logical representations. Because we are only focusing on the verification of models used for diagnosis, there might also be faults occurring that originate from the mapping of data to their logical representations. For verifying the model, we do not need to deal with this topic. We can stay with the abstract representation of real observations for testing.

Finally, we assume that models can be divided into component models and structural models. We further assume that the component models are generally valid and can be used in several systems. This assumption is of particular importance because one argument in favor of model-based diagnosis is its flexibility in adapting to different systems and its model re-use capabilities.

Let us now come up with a definition of the challenge of testing diagnosis models where we have the following information given:

1. A model M for components of given types and their connections.

For testing we want to have the following:

1. A set of systems Σ and for each system $S \in \Sigma$ a model M_S representing the structure, i.e., its components and connections.
2. For each system S, we want to have a set of inputs, i.e., possible observations, and a set of expected diagnoses. Note that observations include inputs and outputs of a system, and control commands (like opening or closing a switch).

Note that the systems, as well as their inputs, must be obtained such that they may lead the diagnosis engine to compute different values. This principle is well-known in testing where testers focus on revealing faults and try to bring an implementation into a state of failure. For stating the problem, we do not rely on automation. Test cases for diagnosis models, and in particular the behavioral part, may be developed manually or using any method for automated test case generation (if possible).

In practice, we might be interested in testing a particular model comprising a structural and behavioral part of a given system. For this variant of the general model testing challenge, we only need to come up with observations and expected diagnoses. In the next section, we discuss generating test cases using the two-bulb example as a use case.

14.3 Use Case

In this section, we use the two-bulb example from Figure 14.1 as a use case for diagnosis model testing. We developed the diagnosis model using the input format of the Clingo[1] theorem prover that relies on the logic programming language Prolog. In Figure 14.3 we see the source code of the model. In Line 1, the ordinary behavior of a battery is given. In case the battery is correctly working (and the predicate nab\1 is true), the battery provides a nominal output at the pow port. In lines 2-4, we formalize the model of a switch. A switch is transferring the power from the in_pow to the out_pow port and vice versa if it is correctly working an on. If the switch is off, there is no power at the output. Similarly, in lines 5-7, we see the behavior model of bulbs. If there is nominal power on the input, and the bulb is working fine, then the bulb is shining. If there is no power, there is also no light. If there is a light, we know that there must be electricity provided.

In lines 8-10, we have the connection model, stating that there is a transfer from one port of a component to another, and their values must be the same. The latter is stated in Line 10. Afterward, we have the structural model of the circuit. First, we define the components of the circuit b, s, l1, l2 for the battery, switch, lamp 1 and lamp 2 respectively. Second, we state the connections between the ports of the components.

For testing the model of the particular two-bulb system, we have to provide test cases comprising observations (which work as the inputs to the model) and the expected diagnoses (which are the expected outputs). For the two bulb example, the position of the switch (on, off), and the state of the

---

[1]see https://potassco.org

```
1.  val(pow(X),nominal) :- type(X,bat), nab(X).
2.  val(out_pow(X),V) :-    type(X,sw), on(X),
                        val(in_pow(X),V), nab(X).
3.  val(in_pow(X),V) :-     type(X,sw), on(X),
                        val(out_pow(X),V), nab(X).
4.  val(out_pow(X),zero) :- type(X,sw), off(X), nab(X).
5.  val(light(X),on) :- type(X,lamp),
                        val(in_pow(X),nominal), nab(X).
6.  val(light(X),off) :-    type(X, lamp),
                        val(in_pow(X),zero), nab(X).
7.  val(in_pow(X), nominal) :-  type(X,lamp),
                        val(light(X),on).
8.  val(X,V) :- conn(X,Y), val(Y,V).
9.  val(Y,V) :- conn(X,Y), val(X,V).
10. :- val(X,V), val(X,W), not V=W.
11. type(b, bat).
12. type(s, sw).
13. type(l1, lamp).
14. type(l2, lamp).
15. conn(in_pow(s), pow(b)).
16. conn(out_pow(s), in_pow(l1)).
17. conn(out_pow(s), in_pow(l2)).
```

Figure 14.3 A model for diagnosis of the two lamp example from Figure 14.1 comprising the behavior of the components (lines 1-7) and connections (lines 8-10), and the structure of the circuit (lines 11-18).

two bulbs regarding light emission (`on`, `off`) serve as the inputs. It is worth noting that the power supply of the battery might also be observed. However, for the initial testing, we only consider those observations where we do not require additional equipment for measurement in practice. Nevertheless, for testing, we may also consider more observations.

When having 3 observations each having a domain comprising 2 values, we finally obtain 8 test cases covering all combinations. We depict this test cases in Table 14.1. Note that the first two test cases (which are highlighted in gray) cover the correct behavior of the system, where the switch is used to turn on and off lamps. Therefore, we see the empty set as the expected diagnosis in the corresponding column. The other test cases formalize an incorrect behavior of the two-bulb circuit.

For testing the model, we run our diagnosis engine `model_diagnose` using the observations of a test case. In Clingo adding observations to models can be simple done via linking the model into a file where we state the

Table 14.1 All eight test cases used to verify the 2-bulb example comprising the used observations and the expected diagnoses. The **P/F** column indicates whether the original model passes ($\sqrt{}$) or fails ($\times$) the test.

	Observations	Expected diagnoses	P/F
1	`on(s). val(light(l1),on).` `val(light(l2,on))`	$\{\{\}\}$	$\sqrt{}$
2	`off(s). val(light(l1),off).` `val(light(l2,off))`	$\{\{\}\}$	$\sqrt{}$
3	`off(s). val(light(l1),on).` `val(light(l2,off))`	$\{\{s,l2\}\}$	$\sqrt{}$
4	`off(s). val(light(l1),off).` `val(light(l2,on))`	$\{\{s,l1\}\}$	$\sqrt{}$
5	`off(s). val(light(l1),on).` `val(light(l2,on))`	$\{\{s\}\}$	$\sqrt{}$
6	`on(s). val(light(l1),on).` `val(light(l2,off))`	$\{\{l2\}\}$	$\sqrt{}$
7	`on(s). val(light(l1),off).` `val(light(l2,on))`	$\{\{l1\}\}$	$\sqrt{}$
8	`on(s). val(light(l1),off).` `val(light(l2,off))`	$\{\{b\},\{s\},\{l1,l2\}\}$	$\sqrt{}$

observations. For the first test case the file `tle_obs1.pl` comprises the following statements:

```
#include two_lamps_example.pl.
on(s).
val(light(l1),on).
val(light(l2),on).
```

The first line includes the model we show in Figure 14.3, which we store in the file `two_lamps_example.pl`. For executing a test case, we run the diagnosis engine in a shell using the following command: `./model_diagnose -f tle_obs1.pl -fault 2`. In this call, we ask for diagnoses comprising up to two components, which we do via setting the parameter `-fault` to 2. Finally, we used a shell script to carry out all test cases. We see the outcome of testing in column `P/F` in Table 14.1. The model passes all tests successfully.

After checking the correctness of diagnosis results obtained when using the model, we wanted to evaluate the quality of the test suite. In software engineering, measures like code coverage or the mutation score are used for this purpose. Estimating code coverage, i.e., the number of rules used to derive a contradiction for diagnosis is difficult because theorem provers

Table 14.2 Running 7 model mutations Mi, where we removed line i in the original model of Figure 14.3, using the 8 test cases from Table 14.1.

	M1	M2	M3	M4	M5	M6	M7
1	√	√	√	√	√	√	√
2	√	√	√	√	√	√	√
3	√	√	√	×	×	√	×
4	√	√	√	×	×	√	×
5	√	√	√	×	√	√	×
6	√	√	√	√	×	√	×
7	√	√	√	√	×	√	×
8	×	×	√	√	×	√	√

usually do not provide this information. Therefore, we focused on mutation testing [2, 12]. The underlying idea is to modify a program and to have a look at whether this modification can be detected by the test suite. The mutation score is defined as the fraction of the detected and all mutations. There are some issues when computing the mutation score, for example, equivalent mutants, i.e., changes of the program that are not changing the behavior.

For languages like Java, there are tools, e.g., [8]. In our case, because of a lack of tools, we only removed rules as modification operators. In particular, we were interested in looking at the consequences to the diagnosis results when removing a rule from a component model. We define a mutant Mi as the original program (from Figure 14.3) where we removed the rule in Line i. In Table 14.2 we find the results obtained for each mutant. We see that there are two mutants M3 and M6 that cannot be detected by any test cases. Hence, the mutation score for our test suite is $\frac{5}{7} = 0.7143$. To clarify the reason behind not having a mutation score of 1.0 we analyzed the corresponding rules of mutant M3 and M6. M3 allows transferring electricity also from the output to the input, which might be appropriate when dealing with other circuits. M6 covers the case where there is zero power on the input. Because there are no other rules allowing to derive zero power, this rule does not provide anything for the reasoning process for this use case and can be removed. Please note that the rule might be introduced again when considering a different use case where we have to deal with zero power at the input.

The question that remains is whether the component models can be used for other systems as well. To verify the corresponding property, i.e., the component models are generally applicable, we have to come up with new systems and apply test case generation again. In this use case, we slightly modified the original two-bulb example. We added another switch in parallel such that both provide the functionality of an or-gate. The lamps have to be

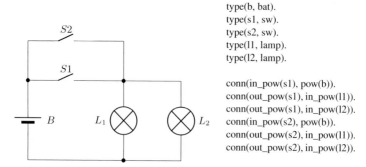

```
type(b, bat).
type(s1, sw).
type(s2, sw).
type(l1, lamp).
type(l2, lamp).

conn(in_pow(s1), pow(b)).
conn(out_pow(s1), in_pow(l1)).
conn(out_pow(s1), in_pow(l2)).
conn(in_pow(s2), pow(b)).
conn(out_pow(s2), in_pow(l1)).
conn(out_pow(s2), in_pow(l2)).
```

Figure 14.4   Another simple electric circuit comprising bulbs, switches and a battery. This circuit is an extended version of the circuit from Figure 14.1. On the right, we have the structural model of this circuit in Prolog notation.

off only if both switches are open, i.e., in their off state. See Figure 14.4 for the schematics of the extended two-bulb circuit.

For testing the extended two-bulb circuit, we have to introduce test cases. Similar to the original circuit, we use all combinations of input values, and manually computed the expected diagnoses. We depict the whole test suite in Table 14.3. There we also see the obtained results after automating the test execution using shell scripts. For many test cases, the computed diagnoses are not equivalent to the expected ones. We conclude that the provided model is not generally applicable.

After carefully analyzing the root cause behind this divergence, we identified the rule in Line 4 of the component model (from Figure 14.3) as problematic. This rule states that an open switch assures that there is no power on the output of the switch. Unfortunately, there might be electricity available because of another power supplying component like given in the extended two-bulb example. Unfortunately, we are also not able to remove this rule because otherwise, the behavior of the original two-bulb example would change (see Table 14.2). A solution would be to introduce a specific or-component that takes the outputs of the two switches as inputs and provides power whenever at least one power output has a nominal value.

14.4 Open Issues and Challenges

We can identify the following results and issues from the use case discussed in the previous section.

Table 14.3 Test cases for the extended two-bulb example from Figure 14.4 and their test execution results. In gray we indicate tests that check the expected (fault-free) behavior of the circuit.

	Observations	Expected diagnoses	P/F
1	on(s1). on(s2). val(light(l1,on)). val(light(l2),on).	$\{\{\}\}$	√
2	off(s1). on(s2). val(light(l1,on)). val(light(l2),on).	$\{\{\}\}$	×
3	on(s1). off(s2). val(light(l1,on)). val(light(l2),on).	$\{\{\}\}$	×
4	off(s1). off(s2). val(light(l1,off)). val(light(l2),off).	$\{\{\}\}$	√
5	on(s1). on(s2). val(light(l1,off)). val(light(l2),on).	$\{\{l1\}\}$	√
6	on(s1). on(s2). val(light(l1,on)). val(light(l2),off).	$\{\{l2\}\}$	√
7	on(s1). on(s2). val(light(l1,off)). val(light(l2),off).	$\{\{b\}, \{s1, s2\}\{l1, l2\}\}$	√
8	on(s1). off(s2). val(light(l1,off)). val(light(l2),on).	$\{\{l1\}\}$	×
9	on(s1). off(s2). val(light(l1,on)). val(light(l2),off).	$\{\{l2\}\}$	×
10	on(s1). off(s2). val(light(l1,off)). val(light(l2),off).	$\{\{b\}, \{s1\}\{l1, l2\}\}$	×
11	off(s1). on(s2). val(light(l1,off)). val(light(l2),on).	$\{\{l1\}\}$	×
12	off(s1). on(s2). val(light(l1,on)). val(light(l2),off).	$\{\{l2\}\}$	×
13	off(s1). on(s2). val(light(l1,off)). val(light(l2),off).	$\{\{b\}, \{s2\}, \{l1, l2\}\}$	×
14	off(s1). off(s2). val(light(l1,on)). val(light(l2),off).	$\{\{s1, s2, l2\}\}$	√
15	off(s1). off(s2). val(light(l1,off)). val(light(l2),on).	$\{\{s1, s2, l1\}\}$	√
16	off(s1). off(s2). val(light(l1,on)). val(light(l2),on).	$\{\{s1, s2\}\}$	√

- *Testing a model* for a particular system that is based on component models and a structural part *is possible* but requires to identify (i) the input, i.e., observations given to the system, and (ii) the expected diagnosis. From this result the following issues arise:

  - We have to identify the observations given to the system. This might not be an obvious task requiring to analyse the functionality of the system. We may start with observations of the input and the output of the system. But this might not be a complete test suite when considering the mutation score.

- Furthermore, we have to consider different observations. We may make use of all combinations as we did in the case study. However, for a larger system, this is infeasible, and other approaches are required. Combinatorial testing [11] might be a good starting point for future research.
- The expected diagnoses have to be computed manually. This is a time-consuming task. Hence, any means for automating this step would be highly appreciated.

- The *generated test suite may not* lead to one that allows for *detecting all faults*. Fault detection capabilities are usually measured using the mutation score. From the use case discussed in the previous section, we see that the mutation score, even when considering only one mutation operator, might be less than 1.0. Related issues and future research activities include:

  - We need to come up with a well-founded theory of mutation testing for logic rules. This also includes considering more mutation operators.
  - There is a need for generating test cases for diagnosis models automatically such that the mutation score can be maximized.

- *Testing should be extended* to check whether the component *models can be used in other systems* as well. What is missing in this context is:

  - The automated generation of different but still relevant systems for practical applications is an open research question. For each of the generated systems, we need to compute test suites and check the correctness of the computed diagnosis. Note that in principle, we have an infinite number of such systems. We have to think about when to stop testing.
  - In case of deviations between the expected diagnoses and the computed ones, someone is interested in identifying the reasons behind them. Hence, we need debugging functionality that may be similar to previous work on debugging knowledge bases [5].

In summary, the main challenge relies on the automation of test case generation. Test cases or at least the expected diagnoses have to be generated manually. Moreover, we need to adapt existing testing methods and techniques for logic representations. Partially there is related work someone can start with. But when compared to corresponding work for ordinary programming languages, available knowledge can be considered minor.

14.5 Conclusion

In this paper, we discussed the use of testing for model-based diagnosis. We focused on assuring the quality of system models used for fault detection and localization. We discussed how to test models and identified arising shortcomings, and future research directions. Testing a system model comes in two flavors: (i) testing a model of a particular system and (ii) testing component models used in different system models. For both, we need to define test cases comprising observations and expected diagnoses. For testing component models, in addition, we need to come up with different systems. Issues and challenges include providing means for answering the question of when to stop testing, giving quality guarantees, and the automation of test case generation.

Acknowledgments

The research was supported by ECSEL JU under the project H2020 826060 AI4DI - Artificial Intelligence for Digitising Industry. AI4DI is funded by the Austrian Federal Ministry of Transport, Innovation, and Technology (BMVIT) under the program "ICT of the Future" between May 2019 and April 2022. More information can be retrieved from `https://iktderzukunft.at/en/` bm🔵ⓥ.

References

[1] A. Beschta, O. Dressler, H. Freitag, M. Montag, and P. Struss. A model-based approach to fault localization in power transmission networks. *Intelligent Systems Engineering*, 1992.

[2] T. Budd, R. DeMillo, R. Lipton, and F. Sayward. Theoretical and empirical studies on using program mutation to test the functional correctness of programs. In *Proc. Seventh ACM Symp. on Princ. of Prog. Lang. (POPL)*. ACM, January 1980.

[3] R. Davis, H. Shrobe, W. Hamscher, K. Wieckert, M. Shirley, and S. Polit. Diagnosis based on structure and function. In *Proceedings AAAI*, pages 137–142, Pittsburgh, August 1982. AAAI Press.

[4] J. de Kleer and B. C. Williams. Diagnosing multiple faults. *Artificial Intelligence*, 32(1):97–130, 1987.

[5] A. Felfernig, G. Friedrich, D. Jannach, and M. Stumptner. Consistency based diagnosis of configuration knowledge bases. In *Proceedings of the*

European Conference on Artificial Intelligence (ECAI), Berlin, August 2000.

[6] G. Friedrich, M. Stumptner, and F. Wotawa. Model-based diagnosis of hardware designs. *Artificial Intelligence*, 111(2):3–39, July 1999.

[7] M. W. Hofbaur, J. Köb, G. Steinbauer, and F. Wotawa. Improving robustness of mobile robots using model-based reasoning. *J. Intell. Robotic Syst.*, 48(1):37–54, 2007.

[8] R. Just. The Major mutation framework: Efficient and scalable mutation analysis for Java. In *Proceedings of the International Symposium on Software Testing and Analysis (ISSTA)*, pages 433–436, San Jose, CA, USA, 2014.

[9] D. Kaufmann, I. Nica, and F. Wotawa. Intelligent agents diagnostics - enhancing cyber-physical systems with self-diagnostic capabilities. *Adv. Intell. Syst.*, 3(5):2000218, 2021.

[10] Y. Koroglu and F. Wotawa. Fully automated compiler testing of a reasoning engine via mutated grammar fuzzing. In *In Proc. of the 14th IEEE/ACM International Workshop on Automation of Software Test (AST)*, Montreal, Canada, 27th May 2019.

[11] D. R. Kuhn, R. N. Kacker, and Y. Lei. *Introduction to Combinatorial Testing*. Chapman & Hall/CRC Innovations in Software Engineering and Software Development Series. Taylor & Francis, 2013.

[12] J. A. Offutt and S. D. Lee. An empirical evaluation of weak mutation. *IEEE Transactions on Software Engineering*, 20(5):337–344, 1994.

[13] C. Picardi, R. Bray, F. Cascio, L. Console, P. Dague, O. Dressler, D. Millet, B. Rehfus, P. Struss, and C. Vallée. Idd: Integrating diagnosis in the design of automotive systems. In *Proceedings of the European Conference on Artificial Intelligence (ECAI)*, pages 628–632, Lyon, France, 2002. IOS Press.

[14] K. Rajan, D. Bernard, G. Dorais, E. Gamble, B. Kanefsky, J. Kurien, W. Millar, N. Muscettola, P. Nayak, N. Rouquette, B. Smith, W. Taylor, and Y.-w. Tung. Remote Agent: An Autonomous Control System for the New Millennium. In *Proceedings of the 14th European Conference on Artificial Intelligence (ECAI)*, Berlin, Germany, August 2000.

[15] R. Reiter. A theory of diagnosis from first principles. *Artificial Intelligence*, 32(1):57–95, 1987.

[16] G. Steinbauer and F. Wotawa. Model-based reasoning for self-adaptive systems - theory and practice. In *Assurances for Self-Adaptive Systems*, volume 7740 of *Lecture Notes in Computer Science*, pages 187–213. Springer, Switzerland, 2013.

[17] F. Wotawa. Testing self-adaptive systems using fault injection and combinatorial testing. In *Proceedings of the Intl. Workshop on Verification and Validation of Adaptive Systems (VVASS 2016)*, pages 305–310, Vienna, Austria, 2016. IEEE.

[18] F. Wotawa. Combining combinatorial testing and metamorphic testing for testing a logic-based non-monotonic reasoning system. In *In Proceedings of the 7th International Workshop on Combinatorial Testing (IWCT) / ICST 2018*, April 13th 2018.

[19] F. Wotawa. On the automation of testing a logic-based diagnosis system. In *In Proceedings of the 13th International Workshop on Testing: Academia-Industry Collaboration, Practice and Research Techniques (TAIC PART) / ICST 2018*, April 9th 2018.

[20] F. Wotawa. Reasoning from first principles for self-adaptive and autonomous systems. In E. Lughofer and M. Sayed-Mouchaweh, editors, *Predictive Maintenance in Dynamic Systems – Advanced Methods, Decision Support Tools and Real-World Applications*. Springer, 2019.

[21] F. Wotawa. Using model-based reasoning for self-adaptive control of smart battery systems. In Moamar Sayed-Mouchaweh, editor, *Artificial Intelligence Techniques for a Scalable Energy Transition – Advanced Methods, Digital Technologies, Decision Support Tools, and Applications*. Springer, 2020.

[22] F. Wotawa and D. Kaufmann. Model-based reasoning using answer set programming. *Applied Intelligence*, 2022.

[23] F. Wotawa, O. A. Tazl, and D. Kaufmann. Automated diagnosis of cyber-physical systems. In *IEA/AIE (2)*, volume 12799 of *Lecture Notes in Computer Science*, pages 441–452. Springer, 2021.

[24] J. M. Zhang, M. Harman, L. Ma, and Y. Liu. Machine learning testing: Survey, landscapes and horizons. *IEEE Transactions on Software Engineering*, 48(1):1–36, 2022.

Index

About the Editors

Ovidiu Vermesan holds a PhD degree in microelectronics and a Master of International Business (MIB) degree. He is Chief Scientist at SINTEF Digital, Oslo, Norway. His research interests are in smart systems integration, mixed-signal embedded electronics, analogue neural networks, edge artificial intelligence and cognitive communication systems. Dr. Vermesan received SINTEF's 2003 award for research excellence for his work on the implementation of a biometric sensor system. He is currently working on projects addressing nanoelectronics, integrated sensor/actuator systems, communication, cyber–physical systems (CPSs) and Industrial Internet of Things (IIoT), with applications in green mobility, energy, autonomous systems, and smart cities. He has authored or co-authored over 100 technical articles, conference/workshop papers and holds several patents. He is actively involved in the activities of European partnership for Key Digital Technologies (KDT). He has coordinated and managed various national, EU and other international projects related to smart sensor systems, integrated electronics, electromobility and intelligent autonomous systems such as E^{3}Car, POLLUX, CASTOR, IoE, MIRANDELA, IoF2020, AUTOPILOT, AutoDrive, ArchitectECA2030, AI4DI, AI4CSM. Dr. Vermesan actively participates in national, Horizon Europe and other international initiatives by coordinating and managing various projects. He is the coordinator of the IoT European Research Cluster (IERC) and a member of the board of the Alliance for Internet of Things Innovation (AIOTI). He is currently the technical co-coordinator of the Artificial Intelligence for Digitising Industry (AI4DI) project.

Franz Wotawa received a M.Sc. in Computer Science (1994) and a PhD in 1996 both from the Vienna University of Technology. He is currently a professor of software engineering at the Graz University of Technology. From the founding of the Institute for Software Technology in 2003 to the year 2009 and starting in 2020 Franz Wotawa has been the head of the institute. His research interests include model-based and qualitative reasoning, theorem proving, mobile robots, verification and validation, and software testing and debugging. Besides theoretical foundations, he has always been interested in closing the gap between research and practice. Starting from October 2017, Franz Wotawa is the head of the Christian Doppler Laboratory for Quality Assurance Methodologies for Autonomous Cyber-Physical Systems. During his career, Franz Wotawa has written more than 430 peer-reviewed papers for journals, books, conferences, and workshops. He supervised 100 master's and 38 Ph.D. students. For his work on diagnosis, he received the Lifetime Achievement Award of the Intl. Diagnosis Community in 2016. Franz Wotawa has been a member of a various number of program committees and organized several workshops and special issues of journals. He is a member of the Academia Europaea, the IEEE Computer Society, ACM, the Austrian Computer Society (OCG), and the Austrian Society for Artificial Intelligence and a Senior Member of the AAAI.

Mario Diaz Nava has a PhD, and M.Sc. both in computer science, from Institut National Polytechnique de Grenoble, France, and B.S. in communications and electronics engineering from Instituto Politecnico National, Mexico. He has worked in STMicroelectronics since 1990. He has occupied different positions (Designer, Architect, Design Manager, Project Leader, Program Manager) in various STMicroelectronics research and development organisations. His selected project experience is related to the specifications and design of communication circuits (ATM, VDSL, Ultra-wideband), digital and analogue design methodologies, system architecture and program management. He currently has the position of ST Grenoble R&D Cooperative Programs Manager, and he has actively participated, for the last five years, in several H2020 IoT projects (ACTIVAGE, IoF2020, Brain-IoT), working in key areas such as Security and Privacy, Smart Farming, IoT System modelling, and edge computing. He is currently leading the ANDANTE project devoted to developing neuromorphic ASICS for efficient AI/ML solutions at the edge. He has published more than 35 articles in these areas. He is currently a member of the Technical Expert Group of the PENTA/Xecs European Eureka cluster and a Chapter chair member of the ECSEL/KDT Strategic Research Innovation Agenda. He is an IEEE member. He participated in the standardisation of several communication technologies in the ATM Forum, ETSI, ANSI and ITU-T standardisation bodies.

Björn Debaillie leads imec's collaborative R&D activities on cutting-edge IoT technologies in imec. As program manager, he is responsible for the operational management across programs and projects, and focusses on strategic collaborations and partnerships, innovation management, and public funding policies. As chief of staff, he is responsible for executive finance and operations management and transformations. Björn coordinates semiconductor-oriented public funded projects and seeds new initiatives on high-speed communications and neuromorphic sensing. He currently leads the 35M€ TEMPO project on neuromorphic hardware technologies, enabling low-power chips for computation-intensive AI applications (www.tempo-ecsel.eu). Björn holds patents and authored international papers published in various journals and conference proceedings. He also received several awards, was elected as IEEE Senior Member and is acting in a wide range of expert boards, technical program committees, and scientific/strategic think tanks.